Springer Theses

Recognizing Outstanding Ph.D. Research

Aims and Scope

The series "Springer Theses" brings together a selection of the very best Ph.D. theses from around the world and across the physical sciences. Nominated and endorsed by two recognized specialists, each published volume has been selected for its scientific excellence and the high impact of its contents for the pertinent field of research. For greater accessibility to non-specialists, the published versions include an extended introduction, as well as a foreword by the student's supervisor explaining the special relevance of the work for the field. As a whole, the series will provide a valuable resource both for newcomers to the research fields described, and for other scientists seeking detailed background information on special questions. Finally, it provides an accredited documentation of the valuable contributions made by today's younger generation of scientists.

Theses may be nominated for publication in this series by heads of department at internationally leading universities or institutes and should fulfill all of the following criteria

- They must be written in good English.
- The topic should fall within the confines of Chemistry, Physics, Earth Sciences, Engineering and related interdisciplinary fields such as Materials, Nanoscience, Chemical Engineering, Complex Systems and Biophysics.
- The work reported in the thesis must represent a significant scientific advance.
- If the thesis includes previously published material, permission to reproduce this must be gained from the respective copyright holder (a maximum 30% of the thesis should be a verbatim reproduction from the author's previous publications).
- They must have been examined and passed during the 12 months prior to nomination.
- Each thesis should include a foreword by the supervisor outlining the significance of its content.
- The theses should have a clearly defined structure including an introduction accessible to new PhD students and scientists not expert in the relevant field.

Indexed by zbMATH.

Chao Yao

Fabrication and Properties of High-Performance 122-Type Iron-Based Superconducting Wires and Tapes

Doctoral Theses accepted by
University of Chinese Academy of Sciences, Beijing, China

Springer

Author
Dr. Chao Yao
Institute of Electrical Engineering
Chinese Academy of Sciences
Beijing, China

Supervisor
Prof. Yanwei Ma
Beijing, China

ISSN 2190-5053 ISSN 2190-5061 (electronic)
Springer Theses
ISBN 978-981-10-5183-8 ISBN 978-981-10-5184-5 (eBook)
https://doi.org/10.1007/978-981-10-5184-5

This Springer imprint is published by the registered company Springer Nature Singapore Pte Ltd.
The registered company address is: 152 Beach Road, #21-01/04 Gateway East, Singapore 189721, Singapore

Supervisor's Foreword

Iron-based superconductors discovered in 2008 formed the second family of high-temperature superconductors next to the cuprates. From the viewpoint of practical applications, the K- or Co-doped 122-type iron-based superconductors ($AeFe_2As_2$, Ae = alkali or alkali earth elements) are very promising for high-field applications, because of their relatively high superconducting transition temperatures, ultra-high upper critical fields, very small anisotropy of superconductivity, and large critical current density of the order of 10^6 A/cm^2 in thin films.

In this thesis, 122-type iron-pnictide superconducting wires and tapes were fabricated based on powder-in-tube method. By using co-doping of metal additives and rolling/pressing-induced grain texture, the transport J_c of 122 tapes was significantly improved. Through magneto-optical imaging technique and high-resolution transmission electron microscopy examination, the grain connectivity and grain boundaries of 122 pnictide tapes were analyzed to understand the mechanism of J_c improvement. In order to further promote the application of iron-pnictide superconductors, Dr. Yao explored the fabrication routes and successfully developed the world's first high-performance multifilamentary 122 pnictide superconducting wire. By hot pressing process, the highest transport J_c has run up to 6.1×10^4 A/cm^2 for 7-filament tapes and 3.5×10^4 A/cm^2 in 19-filament tapes at 4.2 K and 10 T. Moreover, Fe/Ag composite sheathed 114-filament wires and tapes were developed, which is significant for the fabrication of high-strength and low-cost conductors. The excellent results of this thesis demonstrate the promising future for the practical application of iron-based superconductors.

This thesis contains the advanced technologies used to fabricate iron-based superconducting wires and tapes, important methods for characterizing the superconducting properties and microstructure, and also various strategies for enhancing transport J_c performance, which will help the students, researchers, and technical personnel who work in superconducting materials, applied superconductivity, and power industry to obtain a comprehensive understanding of iron-based superconducting wires and tapes.

Beijing, China
May 2022

Prof. Yanwei Ma

Abstract

Iron-based superconductors discovered in 2008 are very promising for high-field applications, because they have relatively high superconducting transition temperatures (T_c), very high upper critical fields (H_{c2}), small anisotropy of superconductivity, and large critical current densities (J_c). In this work, 122-type iron-pnictide superconducting wires and tapes were fabricated based on powder-in-tube (PIT) method. The effects of metal addition, sheath materials, deformation process such as flat rolling, cold pressing, and hot pressing on the performance of the wires and tapes were systematically investigated. The mechanism of J_c improvement was studied by using magneto-optical imaging (MOI), high-resolution transmission electron microscopy (HRTEM) and flux pinning analysis. The fabrication process for multifilamentary iron-based superconducting wires and tapes was explored. The world's first high-performance multifilamentary 122-type iron-pnictide superconducting wire was successfully developed. The main results are listed as follows:

(1) Fe-sheathed $Ba_xK_{1-x}Fe_2As_2$ superconducting tapes were fabricated by the *ex situ* PIT method combined with a short high-temperature annealing technique. The effects of annealing time and different dopants on the transport properties of $Ba_xK_{1-x}Fe_2As_2$ tapes were systematically studied. By co-doping of metal additives Ag and Pb, the transport J_c of $Ba_xK_{1-x}Fe_2As_2$ tape was significantly improved in whole field region and the highest transport J_c was up to 1.4×10^4 A/cm^2 at 4.2 K in self-field. It is proposed that the superior J_c in the co-doped samples are due to the combined effects of Pb doping at low fields and Ag doping at high fields.

(2) By flat rolling and cold pressing processes, highly textured $Sr_xK_{1-x}Fe_2As_2$ tapes with *c*-axis aligned grains were developed based on the PIT method. The effects of deformation parameters such as rolling thickness and deformation pressure on grain texture and transport J_c were systematically investigated. By employing combined rolling-pressing deformation with intermediate heat treatment and hot pressing process, simultaneous optimization of grain connectivity and grain alignment was realized, and thus high transport J_c up to 10^5 A/cm^2 at 4.2 K and 10 T was achieved in 122-type iron-pnictide tapes.

(3) By MOI characterizations, the flux distribution and grain connectivity were systematically characterized for $Sr_xK_{1-x}Fe_2As_2$ tapes. The temperature dependence of the J_c values measured by using the magnetic methods and transport methods is comparatively studied. In the $Sr_xK_{1-x}Fe_2As_2$ tapes, HRTEM characterization on atomic scale revealed that the high J_c can be ascribed to the clean grain boundaries and grains with low misorientation angles. The flux pinning analysis of the high-performance $Sr_xK_{1-x}Fe_2As_2$ tapes showed very high pinning energy. The mechanism of J_c improvement was discussed with respect to grain connectivity, grain texture, and flux pinning.

(4) The fabrication routes and processes for multifilamentary iron-based superconducting wires and tapes were developed based on the PIT method. The relationship between the cold-work deformation process and the superconducting properties was studied in details. After achieving high-performance Fe/Ag sheathed $Sr_xK_{1-x}Fe_2As_2$ 7-filament wires, 19- and 114-filament wires with different sheath configurations were consequently prepared. By hot pressing process, the highest transport J_c has run up to 6.1×10^4 A/cm^2 for 7-filament tapes and 3.5×10^4 A/cm^2 for 19-filament tapes at 4.2 K and 10 T. The transport J_c shows extremely weak field dependence, demonstrating the promising future of 122-type iron-pnictide wires for high-field applications.

Parts of this thesis have been published in the following journal articles:

[1] C. Yao, H. Lin, Q. Zhang, X. Zhang, D. Wang, C. Dong, Y. Ma, S. Awaji, K. Watanabe, Critical current density and microstructural homogeneity in multi-filamentary $Sr_{1-x}K_xFe_2As_2$ composite conductors. J. Appl. Phys. **118**, 203909 (2015).

[2] C. Yao, H. Lin, X. Zhang, C. Dong, D. Wang, Q. Zhang, Y. Ma, S. Awaji, K. Watanabe, Transport critical current density of $Sr_{0.6}K_{0.4}Fe_2As_2$/Ag superconducting tapes processed by flat rolling and uniaxial pressing. IEEE Trans. Appl. Supercond. **25**, 7300204 (2015).

[3] C. Yao, C. Wang, X. Zhang, D. Wang, H. Lin, Q. Zhang, Y. Ma, Y. Tsuchiya, Y. Sun, T. Tamegai, A comparative study of $Sr_{1-x}K_xFe_2As_2$ and $SmFeAsO_{1-x}F_x$ superconducting tapes by magneto-optical imaging. Supercond. Sci. Technol. **27**, 044019 (2014).

[4] X. Zhang, C. Yao, H. Lin, Y. Cai, Z. Chen, J. Li, C. Dong, Q. Zhang, D. Wang, Y. Ma, H. Oguro, S. Awaji, K. Watanabe, Realization of practical level current densities in $Sr_{0.6}K_{0.4}Fe_2As_2$ tape conductors for high-field applications. Appl. Phys. Lett. **104**, 202601.

[5] C. Yao, Y. Ma, X. Zhang, D. Wang, C. Wang, H. Lin, Q. Zhang. Fabrication and transport properties of $Sr_{0.6}K_{0.4}Fe_2As_2$ multifilamentary superconducting wires. Appl. Phys. Lett. **102**, 082602 (2013).

[6] C. Yao, H. Lin, X. Zhang, D. Wang,Q. Zhang, Y. Ma, S. Awaji, K. Watanabe, Microstructure and transport critical current in $Sr_{0.6}K_{0.4}Fe_2As_2$ superconducting tapes prepared by cold pressing. Supercond. Sci. Technol. **26**, 075003 (2013).

[7] Y. Ma, C. Yao, X. Zhang, H. Lin, D. Wang, A. Matsumoto, H. Kumakura, Y. Tsuchiya, Y. Sun, T. Tamegai, Large transport critical currents and magneto-optical imaging of textured $Sr_{1-x}K_xFe_2As_2$ superconducting tapes. Supercond. Sci. Technol. **26**, 035011 (2013).

[8] C. Yao, C. Wang, X. Zhang, L. Wang, Z. Gao, D. Wang, C. Wang, Y. Qi, Y. Ma, S. Awaji, K. Watanabe, Improved transport critical current in Ag and Pb co-doped $Ba_xK_{1-x}Fe_2As_2$ superconducting tapes. Supercond. Sci. Technol. **25**, 035020 (2012).

Acknowledgments

This thesis was completed under the careful guidance and kind help of my supervisor Prof. Yanwei Ma. I am grateful for his profound scientific insight and continuous encouragement, which help me a lot to overcome the difficulties in my research. He also provided us advantageous experimental environment and harmonious researching atmosphere, in which I was inspired and motivated by the fascination and beauty of science.

I appreciate the kind help and selfless support from Prof. Xianping Zhang, Prof. Dongliang Wang, Prof. Zhaoshun Gao, Dr. Lei Wang, Dr. Yanpeng Qi, Dr. Chunlei Wang, Dr. Chengduo Wang, Prof. Chiheng Dong, Dr. Qianjun Zhang, Dr. He Lin, and Dr. Zhiyu Zhang. I always cherish the time in which we work side by side for superconducting wires and tapes.

I would like to thank Prof. Kazuo Watanabe and Prof. Satoshi Awaji in Institute for Materials Research, Tohoku University for their support for high-field critical current measurements; Prof. Tsyoshi Tamegai, Dr. Yuji Tschuiya, and Dr. Yue Sun in the University of Tokyo for their help in the magneto-optical image characterization; Prof. Hiroaki Kumakura in National Institute for Materials Science (NIMS) for the critical current testing under various temperatures; Prof. Laifeng Li and Prof. Rongjin Huang in Technical Institute of Physics and Chemistry, Chinese Academy of Sciences for cold press experiment; Prof. Jianqi Li, Dr. Yao Cai, and Dr. Zhen Chen in Institute of Physics, Chinese Academy of Sciences for experiment using high-resolution transmission electron microscopy.

I would also like to thank Prof. Liangzhen Lin, Prof. Liye Xiao, Prof. Guomin Zhang, Prof. Hongwei Gu, Prof. Qiuliang Wang, Ms. Dongmei Fu, Ms. Hongyan Sun, Ms. Shuyuan Wang, Ms. Xi Xu, and Ms. Weiwei Zhou in Institute of Electrical Engineering, Chinese Academy of Sciences for their kind help for my research and this thesis.

I would like to express my special thanks to my dear wife Xiaoyu Chen.

Finally, I would like to thank my parents Mrs. Lihua Xiang and Mr. Shi Yao for their great support and selfless love.

Contents

Chapter 1
Introduction

For many metals and alloys, when cooled to a sufficiently low temperature, their resistivity suddenly drops to zero. This phenomenon is known as superconductivity, which is first observed by Dutch physicist Heike Kamerlingh Onnes. In 1908, he succeeded in liquefying helium at a temperature of about 4.2 K, and then in 1911, when he measured the low-temperature resistivity of metals, he found the superconductivity of mercury at 4.2 K [1]. After that, researchers observed superconductivity in many other substances, and some of them have higher superconducting transition temperatures. At the same time, due to the exotic nature of superconductors, people have also carried out extensive research for their technical application for electrical engineering, transportation, biomedicine and so on.

1.1 History of Superconducting Materials

After discovering the zero resistance of the superconductor, in 1933 German physicists W. Meissner and R. Ochsenfeld found that if a superconductor was cooled below the transition temperature in the magnetic field, the magnetic field would be completely ejected from the superconductor. This phenomenon is called the Meissner effect [2], which is another essential characteristic of superconductivity. The Meissner effect and the zero resistance are two basic properties for superconductors, which are used to determine the superconducting state for the superconductors. In the early research for superconductors, it was found that the superconducting state is not only related to the temperature, but also to the external magnetic field and the current in the superconductor. When the magnetic field applied to the superconductor is larger than a certain critical value H_c, the superconducting state will be destroyed. When the current passing through a superconductor is higher than a critical current I_c, the superconducting state will also be destroyed, even if the external magnetic field is not applied.

C. Yao, *Fabrication and Properties of High-Performance 122-Type Iron-Based Superconducting Wires and Tapes*, Springer Theses,
https://doi.org/10.1007/978-981-10-5184-5_1

Superconductors can be divided into two categories according to the ways their superconducting state being broke down in applied magnetic field. For the "type-I superconductors", which has only one critical field when superconducting state transforming into normal state, the critical current density and critical field are both too low for practical applications. For the "type-II superconductors", which has a mixed state between the superconducting state and normal state, has two critical fields, the lower critical field H_{c1} and the upper critical field H_{c2}. When in a mixed state, the magnetic flux can run through the areas in normal state in the type-II superconductors while other areas are in superconducting state. The type-II superconductors typically have a higher critical field and higher critical current density, so they are more suitable for practical applications [3].

The earliest study of the mechanism of superconductivity mechanism can be traced back to the 1930s. In 1935, F. London and H. London proposed the so called London equation which tried to explain the zero resistance and Meissner effect in superconductors. By calculation using this equation, they predicted that there is a depth of penetration of magnetic field from the surface of the superconductor, which is now known as the London penetration depth [4]. In 1950, A. Pippard made an important amendment to the London equation, and proposed the concept of coherent length, which is the distance within which the superconducting electron density does not change drastically [5]. In the same year, in order to treat the intermediate state for type-I and the mixed state for type-II superconductors in which magnetic field and superconductivity coexist, V. Ginzberg and L. Landau proposed a theory (G-L theory), which is developed from Landau's theory of second-order phase transitions in 1937 [6]. Although the above-mentioned phenomenological theory has achieved great success in explaining the macroscopic properties of superconductivity, it is impenetrable in microcosmic mechanisms of superconductivity. In 1950s, after the discovery of isotope effect of superconductivity and the proposal of some important concepts including superconducting energy gap, people gradually found the clues to reveal the mechanism of superconductivity. The isotope effect allowed for the assumption that superconductivity is the result of some interaction between electrons and phonons. The interaction between electrons and phonons results in resistance at high temperatures, while at low temperatures it can induce superconductivity. In 1957, J. Bardeen, L. Cooper, and J. Schrieffe proposed a microscopic interpretation of superconductivity, commonly known as BCS theory [7]. At the low temperatures below superconducting transition point, the pairing of electrons close to the Fermi level into Cooper pairs through interaction with the crystal lattice, and these pairs of electrons keep a constant total momentum in the interaction with the crystal lattice, thus the superconductor exhibits zero resistivity.

Since the discovery of superconductivity in mercury, lots of superconducting materials were found, as shown in Fig. 1.1. During the years from 1911 to 1932, lead, tin, niobium and other metal were found to be superconductors, and among them niobium has the highest superconducting transition temperature of 9.2 K. During the period 1932–1973, many alloys, and carbon- and nitrogen-based compounds with superconductivity were discovered, and a high superconducting transition temperature of 23.2 K was observed in Nb_3Ge films. Among these superconducting alloys

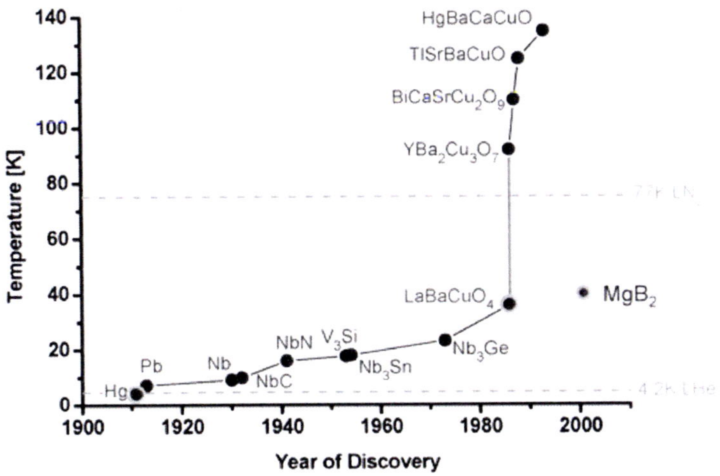

Fig. 1.1 Discovery of superconductors and their superconducting transition temperatures. The points circled in grey mark the discovery of superconductivity in mercury by H. Kamerlingh Onnes in 1911, cuprates by J. Bodnorz and K. Müller in 1986, and MgB_2 by J. Akimitsu in2001. Reproduced from website: www.hoffman.physics.harvard.edu

and intermetallic compounds, NbTi and Nb_3Sn are the most promising ones for practical applications, with a superconducting transition temperature of 9.5 K and 18.1 K, respectively [8, 9]. They have high critical current densities and high upper critical fields, and can be easily made into superconducting wires, which are used to fabricate superconducting magnets operated at the liquid helium temperature 4.2 K. Due to the relatively low superconducting transition temperatures, the metal, alloy and intermetallic compounds are called low-temperature superconductors (LTS). At 4.2 K, Nb–Ti and Nb_3Sn have an upper critical field of 11 and 25 T, respectively. Both of them have current densities over 10^5 A/cm^2, which are about 2 orders of magnitude higher than that of copper conductors, enabling the construction of superconducting magnets that can generate much higher magnetic fields than conventional resistive magnets.

In 1986, J. Bodnorz and K. Müller discovered LaBaCuO superconductors with a superconducting transition temperature of 35 K, which open the gate of searching for high-temperature superconductors (HTS) [10]. In 1987, the superconducting transition temperature in this system was rapidly increased above the liquid nitrogen temperature (77 K) for the first time because of the discovery of $YBa_2Cu_3O_x$ (YBCO) superconductors with transition temperatures up to 93 K [11, 12]. Then Bismuth-based cuprate superconductors including $Bi_2Sr_2CaCu_2O_8$ (Bi-2212) and $Bi_2Sr_2Ca_2Cu_3O_{10}$ (Bi-2223) with transition temperatures up to 110 K were discovered [13]. At present, the highest superconducting transition temperature (138 K at normal pressure and 164 K at high pressure) was found in $Hg_{1-x}Tl_xBa_2Ca_2Cu_3O_{8+\delta}$ system [14]. Bi-2223 and REBCO can carry large supercurrents up to 30–50 K, in field and at 77 K in self-field, so they are promising not only for high field magnets operated in low or moderate temperature region but also for electro-technical applications with much cheaper liquid nitrogen as coolant. On the other hand, Bi-2212 can be used only in low temperature region (<20 K), but it has its own advantages for high-filed applications. Though the transition temperatures of cuprate HTS are much higher than that of LTS, they are hard to be processed into wires due to their ceramic brittleness. In addition, there is weak-linking effect between their grains with high grain boundary angles, which is also not beneficial to the large-scale application.

In 2001, the superconductivity at 39 K in MgB_2 was discovered in Akimitsu group in Aoyama-Gakuin University [15]. The ctitical temperature of MgB_2 is the highest for the bulk binary intermetallic compounds. The origin of superconductivity in MgB_2 can be explained by BCS theory, and is related to the high frequency of vibrations of light boron atoms in the boron layers of the crystal lattice. In contrast to cuprate HTS, the supercurrents in MgB_2 are not sensitive to the weak-linked grain boundaries, so MgB_2 is quite promising for fabricating high-performance wires. However, due to the weak flux pinning ability, the critical current density of MgB_2 drops rapidly with the increase of applied magnetic field, which limits the application of MgB_2 in high magnetic field region. Therefore, MgB_2 is have great potential in applications at around 20 K that can be easily achieved by liquid hydrogen or cryocoolers, and is considered to replace traditional low-temperature superconductors such as Nb–Ti and Nb_3Sn used in liquid helium.

1.2 Iron-Based Superconductors and Their Application Prospects

In Jan. 2008, the Hosono group in Tokyo Institute of Technology discovered the iron-based superconductor (IBS) LaFeAsO, which has a transition temperature (T_c) of 26 K [16]. After that, several different types of iron-based superconductors have also been discovered by other groups. The iron-based superconductors are the second HTS system after the cuprate HTS system, which is of great significance for the research of mechanism of superconductivity, and also the exploration of new superconducting materials. At present, according to the crystal structures of the parent compounds, iron-based superconductors can be divided into several types: (1) 1111-type, including LnOFePn (Ln = La, Ce, Pr, Nd, Sm, Gd, Tb, Dy, Ho, Y; Pn = P, As) and DvFeAsF (Dv = Ca, Sr); (2) 122-type, including AFe_2As_2 (A = Ba, Sr, K, Cs, Ca, Eu) and $A_{1-x}Fe_{2-y}Se_2$ (A = K, Cs, Rb, Tl); (3) 111-type, including AFeAs (A = Li, Na); (4) 11-type, including FeCh (Ch = S, Se, Te); (5) iron-based superconductors with perovskite-like block layers, including 21113- and 32225-types, as shown in Fig. 1.2.

On Mar. 25, 2008, fluorine doped 1111-type IBS $SmO_{1-x}F_xFeAs$ [18] and $CeO_{1-x}F_xFeAs$ [19] with T_c higher than 40 K were reported by X. Chen et al. in University of Science and Technology of China and N. Wang et al. in Institute of Physics, Chinese Academy of Sciences (IOP, CAS), respectively. Then on Mar. 28,

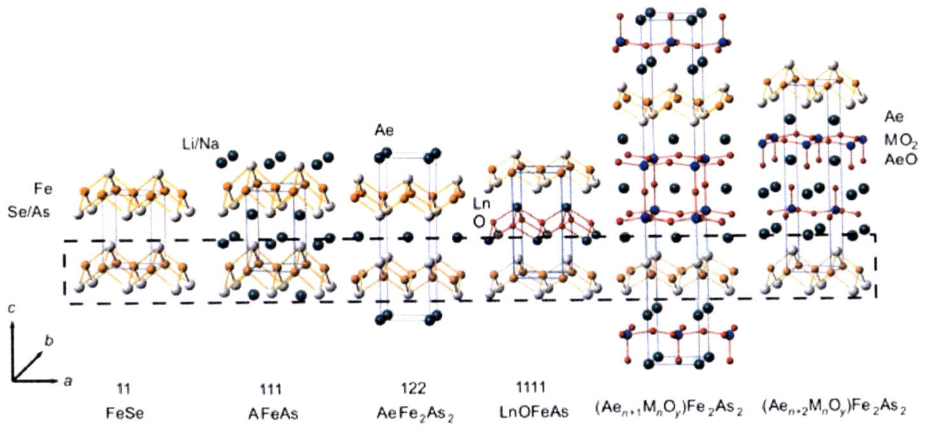

Fig. 1.2 Schematic view of the crystal structures of several typical iron-based superconductor families, in which A, Ae, Ln, and M represent alkali, alkali earth, lanthanide, and transition metal atoms, respectively. Reprinted from Luo and Chen [17]

$PrO_{1-x}F_xFeAs$ superconductors with a T_c of 52 K were reported by Z. Ren et al. in IOP, CAS [20]. They then found that the T_c of 1111-type IBS can be further enhanced to 55 K in $SmO_{1-x}F_xFeAs$ superconductors [21]. At the same time, H. Wen et al. synthesized strontium doped 1111-type IBS $La_{1-x}Sr_xOFeAs$, which is the first hole-doped IBS and has a T_c of 25 K [22]. In June 2008, M. Rotter et al. synthesized a new iron-based parent material $BaFe_2As_2$ in Ludwig-Maximilian University of Munich, and by doping potassium on the barium sites, they obtained the 122-type IBS $Ba_{1-x}K_xFe_2As_2$ with T_c up to 38 K [23]. Soon after that, K. Sasmal at el. in University of Houston synthesized potassium and caesium doped 122-type IBS $Sr_{1-x}K_xFe_2As_2$ and $Sr_{1-x}Cs_xFe_2As_2$ with T_c up to 37 K [24]. In July, M. Wu group in Institute of Physics, Academia Sinica reported the first 11-type IBS $FeSe_{1-x}$ with a T_c of 8 K [25], which has the simplest structure of all IBSs. The 111-type IBS LiFeAs was first synthesized by Wang X. et al. in IOP. CAS [26, 27]. The important results for these four types of IBS are summarized in Table 1.1.

For the four types of iron-based superconductors mentioned above, the highest superconducting transition temperature was observed in 1111-type IBS, which also has very high upper critical fields. As reported by F. Hunte et al., the $H_{c2}(0)$ of $LaFeAsO_{1-x}F_x$ has reached 100 T [28]. For $SmOFeAs_{1-x}F_x$ and $NdOFeAs_{1-x}F_x$ superconductors, the $H_{c2}(0)$ is even higher, estimated to be above 200 T [38–41]. The transition temperature of 122-type IBS is slightly lower than that of the 1111-type, but the superconducting anisotropy is smaller for 122-type IBS, which is more advantageous for magnet fabrication. Based on the experimental results of single crystals, the $H_{c2}(0)$ of $Ba_{1-x}K_xFe_2As_2$ is estimated to be above 100 T [42–45]. As reported by H. Yuan et al., the $H_{c2}(0)$ of $Ba_{1-x}K_xFe_2As_2$ single crystals measured by pulsed magnetic field is 60 T, and the anisotropy of H_{c2} is about 2 at the temperature near its T_c, and drops to about 1 at 5 K [31]. On the other hand, both 1111-type and 122-type IBS have very high critical current density in single crystals. It was

Table 1.1 List of some typical iron-based superconductors

1111-type iron-based superconductors		
LnFePnO (Ln = La, Ce, Pr, Nd, Sm, Gd, Tb, Dy, Ho, Y; Pn = P, As); DvFeAsF (Dv = Ca, Sr)		
Materials	First reported by	Features
$LaFeAsO_{1-x}F_x$	Hosono group in Tokyo Institute of Technology	The first iron-based superconductor [16]
$La_{1-x}Sr_xOFeAs$	H. Wen et al. in Institute of Physics, Chinese Academy of Sciences	The first hole-doped iron-based superconductor [22]
$SmFeAsO_{1-x}F_x$	Z. Ren et al. in Institute of Physics, Chinese Academy of Sciences	$T_c = 55$ K [21]
$LaFeAsO_{1-x}F_x$	F. Hunte et al. in National High Magnetic Field Laboratory, Florida State University	$H_{c2}(0) > 60$ T [28]
$SmFeAsO_{1-x}F_x$	N. Zhigadloin in Swiss Federal Institute of Technology in Zurich	$J_c^{ab} = 2 \times 10^6$ A/cm^2 at 4.2 K [29]
122-type iron-based superconductors		
AFe_2As_2 (A = Ba, Sr, Ca, Eu)		
Materials	First reported by	Features
$Ba_{1-x}K_xFe_2As_2$	M. Rotter et al. in Ludwig-Maximilian University of Munich	The first 122-type IBS with T_c up to 38 K [23]
$Sr_{1-x}K_xFe_2As_2$	K. Sasmal el. in University of Houston	T_c up to 37 K [24]
$SrFe_{2-x}Co_xAs_2$	A. Leithe-Jasperet al. in Max Planck Institute for Chemical Physics of Solids	The first electron doping for 122-type IBS [30]
$Ba_{1-x}K_xFe_2As_2$	H. Yuan et al. in Institute of Physics, Chinese Academy of Sciences	$H_{c2}(0) \sim 60$ T, nearly isotropic superconductivity [31]
$Ba_{1-x}K_xFe_2As_2$	H. Wen et al. in Institute of Physics, Chinese Academy of Sciences	$J_c^{ab} = 5 \times 10^6$ A/cm^2 at 4.2 K [32]
$BaFe_{2-x}Co_xAs_2$	R. Prozorov et al. in Iowa State University; A. Yamamoto et al. in National High Magnetic Field Laboratory, Florida State University	$J_c^{ab} = 2.6 \sim 4 \times 10^6$ A/cm^2 at 4.2 K [33, 34]
$K_xFe_{2-y}Se_2$	J. Guo et al. in Institute of Physics, Chinese Academy of Sciences	$T_c \sim 30$ K [35]

(continued)

reported that the intra-grain J_c of $SmOFeAs_{1-x}F_x$ and $Ba_{1-x}K_xFe_2As_2$ is over 10^6 A/cm^2 [29, 32, 46], and in Ba $(Fe_{1-x}Co_x)_2As_2$ single crystals the intra-grain J_c is also over 10^5 A/cm^2 [33, 34, 47, 48]. Such intra-grain J_c decreases with the increase of external magnetic field below 1 T, and when the applied magnetic field is above 1 T, the J_c shows a very weak field dependence, which can be ascribed to the strong vortex pinning and small anisotropy of IBS. Similar to the results for single crystals, transport J_c exceeding 10^6 A/cm^2 can be obtained in Ba $(Fe_{1-x}Co_x)_2As_2$ epitaxial films [49, 50], which further confirms the superior J_c property of IBS. The 11-type

Table 1.1 (continued)

111-type iron-based superconductors		
AFeAs (A = Li, Na)		
Materials	First reported by	Features
LiFeAs	X. Wang et al. in Institute of Physics, Chinese Academy of Sciences	The first 111-type IBS with T_c up to 18 K [26]
NaFeAs	D. Parker et al. in University of Oxford	$T_c = 9$ K [36]
11-type iron-based superconductors		
FeSe(Te)		
Materials	First reported by	Features
FeSe	F. Hsu et al. in Institute of Physics, Academia Sinica	The first 11-type IBS with $T_c =$ 8 K [25]
FeSe	S. Margadonna et al. in University of Edinburgh	$T_c = 37$ K at ~7 GPa [37]

IBS has a very simple structure, but their T_c is much lower than 1111- and 122-type IBS. The researches on 111-type IBS is not as many as 1111-, 122- and 11-type IBS, whose electromagnetic properties are more attractive for practical applications.

In Table 1.2 we compare the superconducting properties of 1111-, 122- and 11-type IBS with other practical superconducting materials. It can be seen that the T_c of 1111- and 122-type IBS are between the cuprate superconductors (e.g. YBCO, Bi-2212 and Bi-2223) and the traditional low-T_c superconductors (e.g. Nb–Ti and Nb_3Sn). Similar to cuprate superconductors, the H_{c2} of 1111- and 122-type IBS is much higher than low-T_c superconductors. For example, $NdFeAsO_{0.7}F_{0.3}$ and $Ba_{0.55}K_{0.45}Fe_2As_2$ superconductors can be used under high magnetic fields of 40–50 T at 20 K, as shown in Fig. 1.3. Moreover, the anisotropy of IBS is very small, which is preferred for superconducting magnets construction. Therefore, 1111- and

Table 1.2 Basic material parameters relevant for practical superconductors

Material	T_c (K)	$B_{c2,\,4.2\,K}$ (T)	$J_{c,\,4.2\,k}$ (A/cm^2)	Coherence length ε_{ab} (nm)	Anisotropy γ_H
Nb47wt%Ti	0.5	11.5	4×10^5 (5 T)	4	Negligible
Nb_3Sn	18	25	~10^6	3	Negligible
MgB_2	39	18	~10^6	6.5	2 ~ 2.7
YBCO	92	>100	~10^7	1.5	5 ~ 7
Bi-2223	108	>100	~10^6	1.5	50 ~ 90
Bi-2212	90	>100	~10^6	1.5	50 ~ 90
Sm-1111	55	>100	~10^6	1.8 ~ 2.3	4 ~ 5
Ba-122	38	>80	~10^6	1.5 ~ 2.4	1.5 ~ 2
Fe(Se, Te)	16	>40	~10^5	1.2	1.1 ~ 1.9

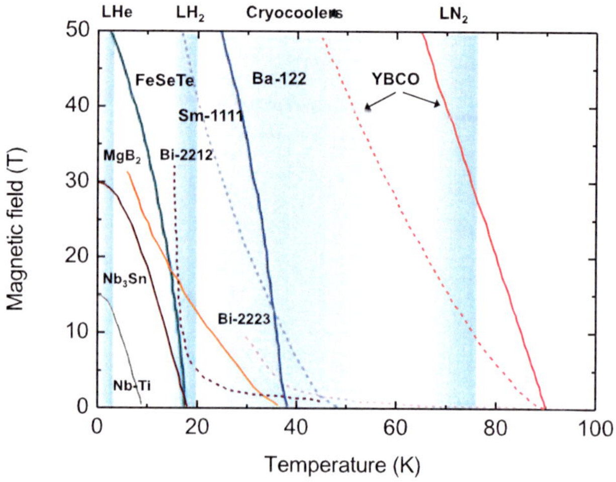

Fig. 1.3 Comparative *H-T* phase diagram for representative cuprates, iron-based superconductors, and conventional superconductors, where the solid and dashed lines show, respectively, $H_{c2}(T)$ and $H^*(T)$ parallel to the c-axis. Reproduced from Yao and Ma [52]

122-type IBS are very promising for high-field application at liquid helium temperature and also moderate temperature region around 20 K, which can be reached by cryocoolers [51].

For practical applications such as superconducting magnets and cables, we must first produce high-performance superconducting wires and tapes with sufficient length including polycrystalline superconducting phase on/in metal substrates/sheaths. Therefore, we should consider not only the basic parameters in IBS single crystals, but also the grain boundary (GB) nature for polycrystals. In cuprate oxide superconductors, the GB is an important issue that greatly affects the transport properties of materials. Due to the weak-link effect at GBs, the intergrain current decrease exponentially with the increase of GB angles in the region of 3° ~ 40° in YBCO bicrystals [53]. To solve this problem, the YBCO conductors are prepared through the epitaxial deposition of YBCO onto a textured template of one or more oxide buffer layers and a normal metal substrate. The template is usually made by introducing texture either into the buffer layers by ion beam-assisted deposition (IBAD) [54], or into the metal substrate using the rolling-assisted biaxially textured substrate (RABiTS) method [55]. At present, the critical current density J_c in these coated conductors is above 10^6 A/cm^2 at 77 K [56]. However, the structure and manufacturing process for the coated conductors are very complex, and the ratio of superconducting layer is also very low for the whole conductors. For these reasons, the manufacture costs and material costs are still too high for the large-scale application using YBCO coated conductors.

In order to study the inter-grain currents for iron-based superconductors, polycrystalline Sm-1111 IBS bulk samples with improved phase purity and mass density were prepared by A. Yamamoto et al. using a hot isostatic pressing (HIP) process. By combined magneto-optical and remanent magnetization analyses on intergranular current density, they concluded that polycrystalline pnictides may have weak-link problems at GBs like polycrystalline cuprate superconductors [62]. The relationship

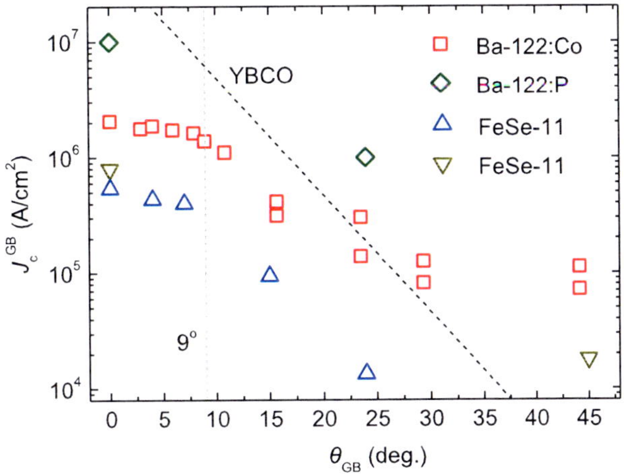

Fig. 1.4 Intergrain transport critical current density J_c^{GB} at in a self-field as a function of misorientation angle θ_{GB} for Co and P doped Ba-122 [57, 58] and FeSe-11 [59, 60] bicrystal GBs. The average data for the YBCO BGB junctions taken at 4 K [61] are also indicated by the black dashed lines for comparison

between the weak-link effect and the grain misorientation angles was studied by S. Lee et al. [63]. They grew epitaxial ~350 nm thick $Ba(Fe_{1-x}Co_x)_2As_2$ thin films on [001] tilt (100) $SrTiO_3$ bicrystal substrates using pulsed laser deposition (PLD). It is confirmed that the critical current density J_c across [001] tilt grain boundaries is depressed. More systematically, T. Katase et al. studied the inter-grain currents across grains with misorientation angles of 3–45° [57]. It is found that the J_c across bicrystal grain boundaries remained high (>1 MA/cm^2) and nearly constant up to a critical misorientation angle of ~9°, which is substantially larger than the ~5° for YBCO. The data measured for FeSe-11 IBS films also show similar results [59, 60]. When misorientation angle is larger than 9°, the intergrain J_c decreases exponentially with the increasing misorientation angle, but slower than that of YBCO, as shown in Fig. 1.4. Therefore, it can be concluded that the high-angle GBs in iron pnictides deteriorate intergrain currents to a lesser extent compared to YBCO cuprate superconductors, and high transport J_c can be expected in iron-based superconducting wires and tapes fabricated by using the simple and low-cost powder-in-tube (PIT) method, which has been used for commercial Nb_3Sn, Bi-2223 and MgB_2 wire products, rather than the complex coated conductor technique for YBCO.

1.3 Progress on Iron-Based Superconducting Wires

Soon after the discovery of iron-based superconductors in 2008, the first attempt to making iron-based superconducting wire was done by Institute of Electrical Engineering, Chinese Academy of Sciences (IEECAS) with an in-situ powder-in-tube (PIT) method, which starts by packing the powders of unreacted precursor materials into a Fe/Ti composite metallic tube. This Fe/Ti sheathed $LaFeAsO_{1-x}F_x$ exhibited superconductivity with a transition temperature (T_c) of 24.6 K [64]. With a similar process, the IEECAS group fabricated a Ta sheathed $SmFeAsO_{1-x}F_x$ wire showing a much higher T_c of 52 K. [65]. Besides these Sm-1111 type iron-based superconducting wires, the first 122-type wire (Nb sheathed $Sr_{1-x}K_xFe_2As_2$ wire) with a T_c of 35.3 K was also reported in 2009 [66]. According to resistivity measurements under magnetic fields, such Sm-1111 and Sr-122 wires have very high upper critical fields (($H_{c2}(0) > 100$ T), indicating great potential in high-field applications.

However, no transport superconducting current was observed when these wires were measured by using the standard four-probe resistive method. By micrographic studies for IBS wires with various sheath materials, it was found that these was a nonsuperconducting layer with a thickness of dozens of micrometers between superconducting cores and Nb, Ta and Fe sheaths. The nonsuperconducting layer was formed by the chemical reaction between the superconducting cores and metallic sheaths during the heat-treatment process for wires, and can subsequently obstruct the transport current passing through the superconducting cores [67]. Fortunately, silver was then found to not react with the superconducting phase, and silver sheath or silver inner sheath combined with other metal as outer sheath have been widely used for 1111- and 122-type IBS wires. By employing Fe/Ag bilayer sheath, transport J_c of 1.2×10^3 and 1.3×10^3 A/cm^2 at 4.2 K in self field was achieved in the $Sr_{1-x}K_xFe_2As_2$ (Sr-122) and $SmFeAsO_{1-x}F_x$ (Sm-1111) wires, respectively [68, 69]. In order to enhance the grain connectivity of superconducting phase, silver and lead as metal additive of an appropriate amount was added in raw materials [68, 70]. After optimizing the starting compounds, the transport J_c of Sm-1111 wires was further increased to 4.6×10^3 A/cm^2 at 4.2 K in self field [71].

After the used of silver sheath, the introduction of ex-situ process instead of the previous in-situ process stimulated a significant improvement of the transport J_c performance of IBS wires. The ex-situ PIT process starts by packing the reacted and well ground and mixed IBS precursor powders into metallic tubes, and can provide a much high mass density and better phase homogeneity for superconducting cores after the final heat treatment of wires. However, the secondary phases at GBs, low density and micro-cracks in IBS phase, which reduce the effective cross section for the current path, still degrade the grain connectivity of superconducting cores. Metal additives such as Ag, Pb and Sn were well mixed with the precursor powders to improve the grain connectivity of IBS wires and tapes, as shown in Fig. 1.5. The first report for ex-situ PIT IBS wires was from the IEECAS group in 2010. With Pb addition to improve the grain connectivity, a Fe/Ag sheathed Sr-122 wire prepared using ex-situ PIT process showed a superior transport J_c of 3.75×10^3 A/cm^2 at 4.2 K in self field [72]. With Ag addition and the ex-situ PIT process, Togano et al. in National Institute for Materials Science (NIMS) and Ding et al. in The University of Tokyo fabricated Ag sheathed Ba-122 wires with transport J_cs on the order of 10^4 A/cm^2 at 4.2 K in self field [73, 74]. For Sm-1111 superconductor, it is hard to obtain high purity precursor due to a relatively higher sintering temperature. On the other hand, the loss of doped fluorine element during sintering process will cause a serious suppression of superconductivity. In 2011, Fujioka et al. in NIMS prepared Sm-1111 wires with ex-situ PIT method. By compensating the fluorine for the reacted Sm-1111 precursor, a transport J_c of about 4.0×10^3 A/cm^2 was measured at 4.2 K in self field [75]. By employing a short-time sintering to reduce the loss of fluorine and

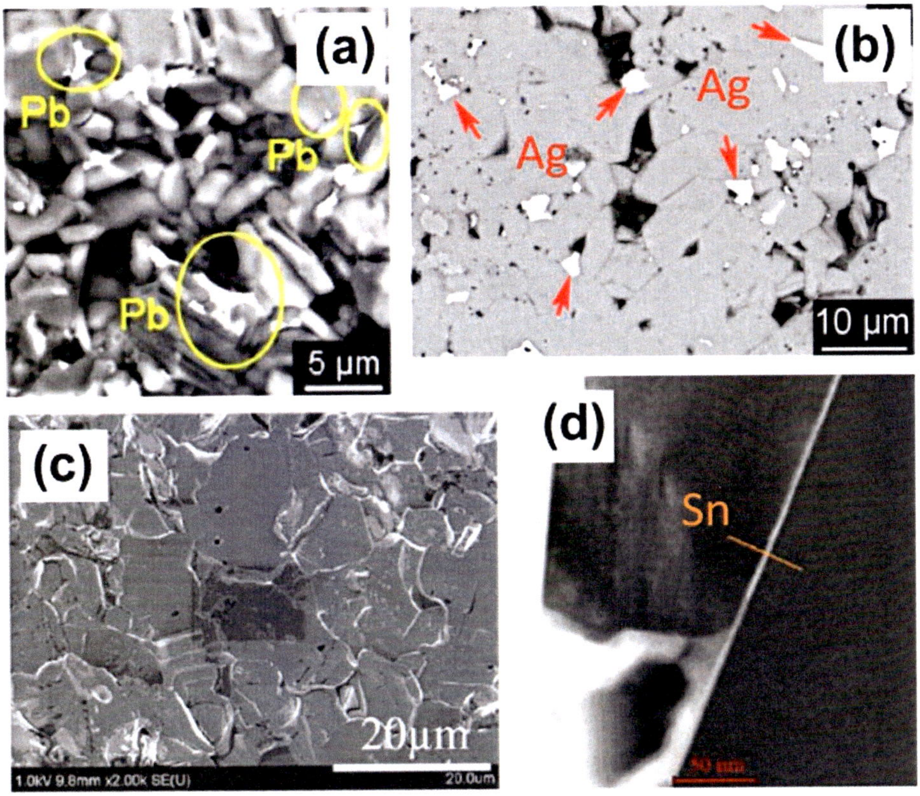

Fig. 1.5 Microstructure of IBS polycrystalline samples with metal addition. Electron backscatter images of **a** Pb-added and **b** Ag-added samples obtained by using a scanning electron microscope (SEM) show that Ag and Pb additives are inhomogenously distributed between IBS grains [70, 78]. **c** A typical SEM images the superconducting core of Sr-122 IBS tapes with Sn addition exhibits no obvious Sn phase [79]. **d** A STEM image showing a Sn-rich layer in 2–3 nm thick between two Sr-122 grains [80]

Sn-addtion to suppress the FeAs impurity phase, the IEECAS group achieved high $J_c s$ above 2×10^4 A/cm^2 at 4.2 K in self field for Sm-1111 wires [76, 77].

Based on the ex-situ PIT process, the development of densification and grain texture techniques for the superconducting phase led to a dramatic increase for the transport J_c of 122-type IBS wires in the following years. In contrast, to the impurity phases in precursors and the difficulty in controlling oxygen and fluorine content in wire annealing process, the progress on J_c improvement for 1111-IBS wires was not as rapid as that for 122-IBS wires, especially for the J_c performance in high-field region. In 2011, the IEEECAS group observed rolling induced texture after as-drawn Fe-sheathed Sr-122 IBS wires being rolled into flat tapes. Since the sheath of these tapes is iron, they were sintered at a high temperature of 1100 °C for a quite short time of 5 ~ 15 min to avoid the reaction between the sheath and superconducting core. As shown in Fig. 1.6, there is a very strong (002) peak in the XRD pattern, indicating *c*-axis texture for Sr-122 grains, which is in accordance with the aligned grains along

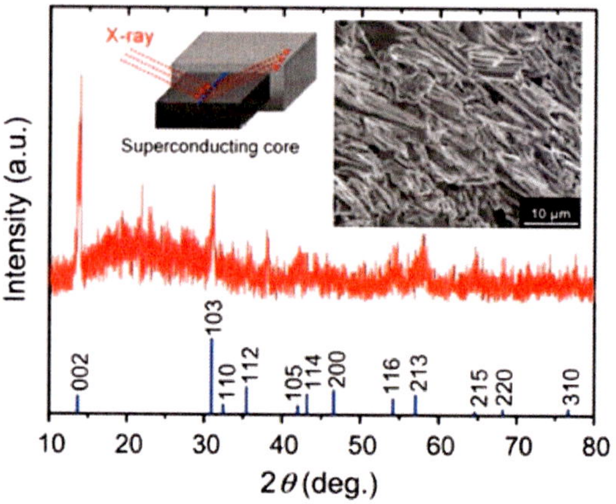

Fig. 1.6 XRD patterns of the superconducting cores for the (Sr, K)-122 tapes produced by flat rolling and annealing at 1100 °C for 5 min. The inset shows the SEM image for the textured (Sr, K)-122 grains. Reproduced from Wang et al. [68]

the rolling direction in the SEM image. Such grain alignment can help to alleviate the weak-link effect at grain boundaries, thus the transport J_c of Pb added Sr-122 IBS tapes was increased to 5.4×10^3 A/cm^2 at 4.2 K in self field [81]. By using Sn instead of Pb additive, the transport J_c of c-axis textured tapes was significantly increased to 2.5×10^4 and 3.5×10^3 A/cm^2 at 4.2 K in self field and under 10 T, respectively [82]. By introducing an optimized annealing process, they obtain a high transport J_c of 1.7×10^4 A/cm^2 at 4.2 K and 10 T in Sn added Sr-122 tapes [80]. This is for the first time that the transport J_c of IBS wires and tapes break through 10^4 A/cm^2 at such a high field of 10 T, indicating the great potential of iron-based superconductors in high-field applications.

Besides the weak-link effect induced by mismatched grains, the voids and cracks in superconducting phase also restrict transport currents. As mentioned above, some metal additives such as Pb, Ag and Sn can enhance the grain connectivity by improving the metallic character of secondary phases, bonding the grains that separated by voids and cracks, and reducing FeAs wetting phase at grain boundaries. On the other hand, enhancing the density of superconducting phase can be another effective way to diminishing the voids for improving the transport J_c performance. In 2012, Ba-122 bulks were prepared by Weiss et al. in Florida State University using a hot isostatic press technique (HIP) to obtain high mass density and at low-temperature to reduce FeAs wetting phase [83]. As presented in Fig. 1.7, the powder X-ray patterns show that small FeAs peaks can be seen from the impurity phase that occupies less than 3% of the sample volume. The TEM image as shown in the inset of Fig. 1.7 revealed that the bulk material consist of equiaxed grains with average grain diameter of ~200 nm, much smaller than the ones in the inset of Fig. 1.6. A selected-area electron diffraction image of TEM indicates that the grains of the material are randomly oriented with many high-angle GBs. With HIP process under a 192 MPa isostatic pressure at 600 °C, a transport J_c of 8.5×10^3 A/cm^2 at 4.2 K and 10 T was achieved in Cu/Ag sheathed Ba-122 wires, in which the mass density of the superconducting core of Ba-122 round wire can be increased from 70 ~ 80% to almost 100%. Though round wires are more attractive than flat tapes for high-field

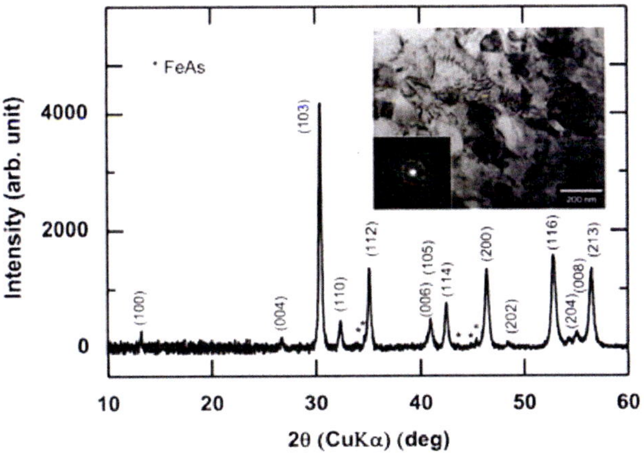

Fig. 1.7 Powder X-ray patterns of (Ba, K)-122 bulk material processed by hot isostatic pressing under 192 MPa of pressure at 600 °C. Small FeAs peaks can be seen from the impurity phase. Inset is the TEM image of the bulk material showing several equiaxed grains with average grain diameter of ~200 nm. A selected-area electron diffraction image indicates that the grains of the material are randomly oriented with many high-angle grain boundaries. Reproduced from Weiss et al. [83]

applications due to their shape isotropies, it is more difficult to introduce grain texture like for flat tapes. This may be the reason why the transport J_c of HIP round wire is lower than that of rolled tapes even with such high mass density.

1.4 Challenges in the Research and Development of 122-IBS Wires

According to the review for iron-based superconducting wires in Sect. 1.3, as summaried in Table 1.3, the remarkable improvements in the transport J_c of 122-IBS wires greatly encourage us to make them practical in the future. For practical applications, a high transport J_c is still the primary challenge in the research and development of wires. Therefore, more efforts are needed to reveal the current-limit mechanism and improve the fabricating process for 122-IBS wires.

(1) Through the studies on the microstructure of 122-IBS wires, the defects in superconducting phase such as micro-cracks, voids, element inhomogeneity, and impurity phase still restricted the transport currents. New fabricating methods or improvements on the currently used process are required to further reduce the impurities, increase the mass density and enhance the grain connectivity of superconducting phase.

(2) Concerning the weak-link behavior at the grain boundaries of mismatched grains, though *c*-axis grain alignment can be induced by flat rolling process, the extent of grain texture is needed to be further increased, for which the

Table 1.3 Summary of 122-type iron-based superconducting wires and tapes

Sample	Process	Transport J_c	Features	References
Nb/Sr-122 wire	In-situ PIT swage + draw 850 °C, 35 h	–	The first 122-type IBS wire	Qi et al. [66]
Fe/Ag/Sr-122 wire and tape	In-situ PIT swage + draw + flat roll (850–900 °C, 35 h)	1.2 kA/cm^2 (4.2 K, 0 T)	Ag used as sheath for the first time; Ag used as metal additive to improve the grain connectivity	Wang et al. [68]
Fe/Ag/Sr-122 tape	In-situ PIT swage + draw + flat roll 850–900 °C, 35 h	1.1 kA/cm^2 (4.2 K, 0 T)	Pb used as metal additive	Wang et al. [70]
Fe/Ag/Sr-122 wire	Ex-situ PIT swage + draw 900 °C, 20 h	3.75 kA/cm^2 (4.2 K, 0 T)	ex-situ PIT process combined with Pb or Ag addition	Qi et al. [72]
Ag/Ba-122 wire	Ex-situ PIT groove roll + swage 850 °C, 30 h	10 kA/cm^2 (4.2 K, 0 T) 1.1 kA/cm^2 (4.2 K, 10 T)	ex-situ PIT process combined with Ag addition	Togano et al. [73]
Fe/Sr-122 tape	Ex-situ PIT swage + draw + flat roll 1100 °C, 5 min	5.4 kA/cm^2 (4.2 K, 0 T)	Rolling induced texture combined with Pb addition	Wang et al. [81]
Fe/Sr-122 tape	Ex-situ PIT swage + draw + flat roll 1100 °C, 0.5–15 min	25 kA/cm^2 (4.2 K, 0 T) 3.5 kA/cm^2 (4.2 K, 10 T)	Rolling induced texture combined with Sn addition	Gao et al. [82]
Cu/Ag/Ba-122 wire	Ex-situ PIT groove roll + draw 600 °C, 192 MPa, 10 h	120 kA/cm^2 (4.2 K, 0 T) ~8.5 kA/cm^2 (4.2 K, 10 T)	Hot isostatic pressing used to achieve high mass density for Ba-122 phase	Weiss et al. [83]
Fe/Sr-122 tape	Ex-situ PIT swage + draw + flat roll 850–900 °C, 1–30 min	17 kA/cm^2 (4.2 K, 10 T)	Textured tape combined with Sn addition by optimized heat treatment	Gao et al. [80]

effects of sheath materials, deformation process and heat treatment process on the grain alignment should be clarified.

(3) Besides the magnetic field dependence of transport J_c, the temperature dependence of transport property at the liquid helium temperature 4.2–30 K should

be investigated. The characterizations of the microstructure of grain boundaries and the flux pinning in superconducting phase for 122-IBS can help to understand the mechanism of J_c improvement in the future.

(4) In order to further promote the practical research of IBS wires, based on the improvement of J_c performance for monofilamentary PIT process, the fabricating route for multifilamentary wires is needed to be explored, since practical applications, especially the ones in high-field regions, require multifilamentary wires for safety reasons.

Focusing on the fabrication for high-performance 122 IBS wires, this thesis will present my works on improving grain connectivity and grain texture for J_c enhancement by utilizing metallic additives, mechanical deformation such as flat rolling and uniaxial pressing, analyzing the superconducting properties of wires by magneto-optical imaging (MOI) technique, high resolution transmission electron microscopy (HRTEM) and flux pinning calculation, and finally developing the multifilamentary fabrication process for IBS wires.

References

1. Kamerlingh Onnes H (1911) The superconductivity of mercury. Comm Phys Lab Univ Leiden 122:122–124
2. Meissner W, Ochsenfeld R (1933) Ein neuer Effekt bei Eintritt der Supraleitfähigkeit. Naturwiss 21:787
3. Tinkham M (1996) Introduction to superconductivity, 2nd edn. McGraw-Hill, Inc
4. London F, London H (1935) The electromagnetic equations of the supraconductor. Proc Roy Soc A 149:71
5. Pippard AB (1950) An experimental and theoretical study of the relation between magnetic field and current in a superconductor. Proc Roy Soc A 216:547
6. Ginzberg V, Landau L (1950) On the theory of superconductivity. Soviet Phys JETP 20:1064
7. Bardeen J, Cooper LN, Schrieffer JR (1957) Theory of superconductivity. Phys Rev 108:1175
8. Kunzler JE, Wernick JH, Hsu FSL, Buehler E (1961) Superconductivity in Nb_3Sn at high current density in a magnetic field of 88 K Gauss. Phys Rev Lett 6:89
9. Polak M, Hlasnik I, Krempask L (1973) Voltage-current characteristics of NbTi and Nb_3Sn superconductors in flux creep region. Cryogenics 13:702
10. Bednorz JG, Muller KA (1986) Possible high T_c superconductivity in the Ba–La–Cu–O system. Z Phys B 64:189
11. Wu MK, Ashburn JR, Torng CJ, Hor PH, Meng RL, Gao L, Huang ZJ, Wang YQ, Chu CW (1987) Superconductivity at 93 K in a new mixed-phase Y–Ba–Cu–O compond system at ambient pressure. Phys Rev Lett 58:908
12. Zhao ZX, Chen LQ, Cui CG et al (1987) Superconductivity above liquid nitrogen temprature in new oxide system. Sci Bull 32:1098
13. Michel C, Hervieu M, Borel MM et al (1987) Superconductivity in the Bi–Sr–Cu–O system. Zeitschrift fur Physik B 68:421
14. Schilling A, Cantoni M, Guo JD et al (1993) Superconductivity above 130 K in the Hg–Ba–Ca–Cu–O system. Nature 363:56–58
15. Nagamatsu J, Nakagawa N, Muranaka T, Zenitani Y, Akimitsu J (2001) Superconductivity at 39 K in Magnesium diboride. Nature 410:63
16. Kamihara Y, Watanabe T, Hirano M, Hosono H (2008) Iron-based layered superconductor $La[O_{1-x}F_x]FeAs$ ($x = 0.05$–0.12) with $T_c = 26$ K. J Am Chem Soc 130:3296

17. Luo X, Chen X (2015) Crystal structure and phase diagrams of iron-based superconductors. Sci China Mater 58:77–89
18. Chen XH, Wu T, Wu G, Liu RH, Chen H, Fang DF (2008) Superconductivity at 43 K in $SmFeAsO_{1-x}F_x$. Nature 453:761
19. Chen GF, Li Z, Wu D, Li G, Hu WZ, Dong J, Zheng P, Luo JL, Wang NL (2008) Superconductivity at 41 K and its competition with spin-density-wave instability in layered $CeO_{1-x}F_xFeAs$. Phys Rev Lett 100:247002
20. Ren ZA, Che GC, Dong XL, Yang J, Lu W, Yi W, Shen XL, Li ZC, Sun LL, Zhou F, Zhao ZX (2008) Superconductivity and phase diagram in iron-based arsenic-oxides $ReFeAsO_{1-\delta}$ (Re = rare-earth metal) without fluorine doping. Europhys Lett 83:17002
21. Ren ZA, Lu W, Yang J, Yi W, Shen XL, Li ZC, Che GC, Dong XL, Sun LL, Zhou F, Zhao ZX (2008) Superconductivity at 55 K in iron-based F-doped layered quaternary compound $SmFeAsO_{1-x}F_x$. Chin Phys Lett 25:2215
22. Wen HH, Mu G, Fang L, Yang H, Zhu XY (2008) Superconductivity at 25 K in hole-doped $(La_{1-x}Sr_x)$ OFeAs. Europhys Lett 82:17009
23. Rotter M, Tegel M, Johrendt D (2008) Superconductivity at 38 K in the iron arsenide $Ba_{1-x}K_xFe_2As_2$. Phys Rev Lett 101:107006
24. Sasmal K, Lv B, Lorenz B, Guloy A, Chen F, Xue Y, Chu CW (2008) Superconducting Fe-based compounds $(A_{1-x}Sr_x)Fe_2As_2$ with A = K and Cs with transition temperatures up to 37 K. Phys Rev Lett 101:107007
25. Hsu FC, Luo JY, Yeh KW, Chen TK, Huang TW, Wu PW, Lee YC, Huang YL, Chu YY, Yan DC, Wu MK (2008) Superconductivity in the PbO-type structure alpha-FeSe. Proc Natl Acad Sci 105:14262
26. Wang XC, Liu QQ, Lv YX, Gao WB, Yang LX, Yu RC, Li FY, Jin CQ (2008) The superconductivity at 18 K in LiFeAs system. Solid State Commun 148:538
27. Tapp JH, Tang ZJ, Lv B, Sasmal K, Lorenz B, Chu CW, Guloy AM (2008) LiFeAs: An intrinsic FeAs-based superconductor with $T_c = 18$ K. Phys Rev B 78:060505
28. Hunte F, Jaroszynski J, Gurevich A, Larbalestier DC, Jin R, Sefat AS, McGuire MA, Sales BC, Christen DK, Mandrus D (2008) Two-band superconductivity in $LaFeAsO_{0.89}F_{0.11}$ at very high magnetic fields. Nature 453, 903
29. Zhigadlo ND, Katrych S, Bukowski Z, Weyeneth S, Puzniak R, Karpinski J (2008) Single crystals of superconducting $SmFeAsO_{1-x}$ F_y grown at high pressure. J Phys: Condens Matter 20:342202
30. Leithe-Jasper A, Schnelle W, Geibel C, Rosner H (2008) Superconducting state in $SrFe_{2-x}Co_xAs_2$ by internal doping of the iron arsenide layers. Phys Rev Lett 101:207004
31. Yuan HQ, Singleton J, Balakirev FF, Baily SA, Chen GF, Luo JL, Wang NL (2008) Nearly isotropic superconductivity in (Ba, K)Fe_2As_2. Nature 457:565
32. Yang H, Luo Q, Wang ZS, Wen HH (2008) Fishtail effect and the votex phase diagram of single crystasl $Ba_{0.6}K_{0.4}Fe_2As_2$. Appl Phys Lett 93:142506
33. Prozorov R, Ni N, Tanatar MA, Kogan VG, Gordon RT, Martin C, Blomberg EC, Prommapan P, Yan JQ, Bud'ko SL, Canfield PC (2008) Vortex phase diagram of $Ba(Fe_{0.93}Co_{0.07})_2As_2$ single crystals. Phys Rev B 78:224506
34. Yamamoto A, Jaroszynski J, Tarantini C, Balicas L, Jiang J, Gurevich A, Larbalestier DC, Jin R, Sefat AS, McGuire MA, Sales BC, Christen DK, Mandrus D (2009) Small anisotropy, weak thermal fluctuations, and high field superconductivity in Co-doped iron pnictide $Ba(Fe_{1-x}Co_x)_2As_2$. Appl Phys Lett 94:062511
35. Guo JG, Jin SF, Wang G, Wang SC, Zhu KX, Zhou TT, He M, Chen XL (2010) Superconductivity in the iron selenide $K_xFe_2Se_2$ $(0 \leq x \leq 1.0)$. Phys Rev B 82:180520
36. Parker DR, Pitcher MJ, Baker PJ, Franke I, Lancaster T, Blundell SJ, Clarke SJ (2009) Structure, antiferromagnetism and superconductivity of the layered iron arsenide NaFeAs. Chem Commun 2009:2189
37. Margadonna S, Takabayashi Y, Ohishi Y, Mizuguchi Y, Takano Y, Kagayama T, Nakagawa T, Takata M, Prassides K (2009) Pressure evolution of the low-temperature crystal structure and bonding of the superconductor FeSe ($T_c = 37$ K). Phys Rev B 80:064506

38. Senatore C, Cantoni M, Wu G, Liu RH, Chen XH, Flukiger R (2008) Upper critical fields well above 100 T for the superconductor $SmFeAsO_{0.85}F_{0.15}$ with $T_c = 46$ K. Phys Rev B 78:054514
39. Wang XL, Ghorbani SR, Peleckis G, Dou S (2009) Very high critical field and superior J_c-field performance in $NdFeAsO_{0.82}F_{0.18}$ with T_c of 51 K. Adv Mater 21:236
40. Jia Y, Cheng P, Fang L, Luo HQ, Yang H, Ren C, Shan L, Gu CZ, Wen HH (2008) Critical fields and anisotropy of $NdFeAsO_{0.82}F_{0.18}$ single crystals. Appl Phys Lett 93:032503
41. Jaroszynski J, Hunte F, Balicas L, Jo YJ, Raičević I, Gurevich A, Larbalestier DC, Balakirev FF, Fang L, Cheng P, Jia Y, Wen HH (2008) Upper cirtical fields and thermally-activated transport of $NdFeAsO_{0.7}F_{0.3}$ single crystal. Phys Rev B 78:174523
42. Ivanovskii AL (2008) New high-temperature superconductors based on rare-earth and transition metal oxyarsenides and related phases: synthesis, properties and simulations. Phys Usp 51:1229
43. Ni N, Bud'ko SL, Kreyssig A, Nandi S, Rustan GE, Goldman AI, Gupta S, Corbett JD, Kracher A, Canfield PC (2008) Anisotropic thermodynamic and transport properties of single crystalline $(Ba_{1-x}K_x)\ Fe_2As_2$ ($x = 0$ and 0.45). Phys Rev B 78:014507
44. Wang XL, Ghorbani SR, Lee SI, Dou SX, Lin CT, Johansen TH, Cheng ZX, Peleckis G, Muller K, Shabazi M, Sun GL, Sun DL (2010) Very strong intrinsic supercurrent carrying ability and vortex avalanches in $(Ba, K)Fe_2As_2$ superconducting single crystals. Phys Rev B 82:024525
45. Wang CL, Gao ZS, Yao C, Wang L, Qi YP, Wang DL, Zhang XP, Ma YW (2011) One-step method to grow $Ba_{0.6}K_{0.4}Fe_2As_2$ single crystals without fluxing agent. Supercond Sci Technol 24:065002
46. Moll PJW, Puzniak R, Balakirev F, Rogacki K, Karpinski J, Zhigadlo ND, Batlogg B (2010) High magnetic-field scales and critical currents in SmFeAs(O, F) crystals. Nature Mater 9:628
47. Ni N, Tillman ME, Yan JQ, Kracher A, Hannahs ST, Bud'ko SL, Canfield PC (2008) Effects of Co substitution on thermodynamic and transport properties and anisotropic H_{c2} in $Ba(Fe_{1-x}Co_x)_2As_2$ single crystals. Phys Rev B 78:214515
48. Shahbazi M, Wang XL, Ghorbani SR, Dou SX, Choi KY (2012) Angular dependence of pinning potential, upper critical field, and irreversibility field in underdoped $BaFe_{1.9}Co_{0.1}As_2$ single crystal. Appl Phys Lett 100:102601
49. Lee S, Jiang J, Zhang Y, Bark CW, Weiss JD, Tarantini C, Nelson CT, Jang HW, Folkman CM, Baek SH, Polyanskii A, Abraimov D, Yamamoto A, Park JW, Pan XQ, Hellstrom EE, Larbalestier DC, Eom CB (2010) Template engineering of Co-doped $BaFe_2As_2$ single-crystal thin films. Nat Mater 9:397
50. Mohan S, Taen T, Yagyuda H, Nakajima Y, Tamegai T, Katase T, Hiramatsu H, Hosono H (2010) Transport and magnetic properties of Co-doped $BaFe_2As_2$ epitaxial thin films grown on MgO substrate. Supercond Sci Technol 23:105016
51. Gurevich A (2014) Challenges and opportunities for applications of unconventional superconductors. Annu Rev Condens Matter Phys 5:35
52. Yao C, Ma Y (2021) Superconducting materials: challenges and opportunities for large-scale applications. iScience 24:102541
53. Gurevich A, Pashitskii EA (1998) Current transport through low-angle grain boundaries in high-temperature superconductors. Phys Rev B 57:13878
54. Iijima T, Tanabe N, Kohno O, Ikeno Y (1992) In-plane aligned $YBa_2Cu_3O_{7-x}$ thin films deposited on polycrystalline metallic substrates. Appl Phys Lett 60:769
55. Goyal A, Norton DP, Christen DK, Specht ED, Paranthaman M, Kroeger DM, Budai JD, He Q, List FA, Feenstra R, Kerchner HR, Lee DF, Hatfield E, Martin PM, Mathis J, Park C (1996) Epitaxial superconductors on rolling-assisted biaxially-textured substrates (RABiTS): a route towards high critical current density wire. Appl Supercond 4:403
56. Yamada Y, Miyata S, Yoshizumi M, Fukushima H, Ibi A, Kinoshita A, Izumi T, Shiohara Y, Kato T, Hirayama T (2009) Development of long length IBAD-MgO and PLD coated conductors. IEEE Trans Appl Supercond 19:3236
57. Katase T, Ishimaru Y, Tsukamoto A, Hiramatsu H, Kamiya T, Tanabe K, Hosono H (2011) Advantageous grain boundaries in iron-pnictide superconductors. Nat Commun 2:409
58. Sakagamia A, Kawaguchia T, Tabuchib M, Ujiharac T, Takeda Y, Ikuta H (2013) Critical current density and grain boundary property of $BaFe_2$(As, P)2 thin films. Physica C 494:181

59. Si W, Zhang C, Shi X, Ozaki T, Jaroszynski J, Li Q (2015) Grain boundary junctions of $FeSe_{0.5}Te_{0.5}$ thin films on $SrTiO_3$ bi-crystal substrates. Appl Phys Lett 106:032602
60. Sarnelli E, Adamo M, Nappi C, Braccini V, Kawale S, Bellingeri E, Ferdeghini C (2014) Properties of high-angle Fe(Se, Te) bicrystal grain boundary junctions. Appl Phys Lett 104:162601
61. Hilgenkamp H, Mannhart J (2002) Grain boundaries in high-T_c superconductors. Rev Mod Phys 74:485–549
62. Yamamoto A, Jiang J, Kametani F, Polyanskii A, Hellstrom E, Larbalestier D, Martinelli A, Palenzona A, Tropeano M, Putti M (2011) Evidence for electromagnetic granularity in polycrystalline Sm1111 iron-pnictides with enhanced phase purity. Supercond Sci Technol 24:045010
63. Lee S, Jiang J, Weiss JD, Folkman CM, Bark CW, Tarantini C, Xu A, Abraimov D, Polyanskii A, Nelson CT, Zhang Y, Baek SH, Jang HW, Yamamoto A, Kametani F, Pan XQ, Hellstrom EE, Gurevich A, Eom CB, Larbalestier DC (2009) Weak-link behavior of gran boundaries in superconducting $Ba(Fe_{1-x}Co_x)_2As_2$ bicrystals. Appl Phys Lett 95:212505
64. Gao Z, Wang L, Qi Y, Wang D, Zhang X, Ma Y (2008) Preparation of $LaFeAsO_{0.9}F_{0.1}$ wires by the powder-in-tube method. Supercond Sci Technol 21:105024
65. Gao Z, Wang L, Qi Y, Wang D, Zhang X, Ma Y, Yang H, Wen H (2008) Superconducting properties of granular $SmFeAsO_{1-x}F_x$ wires with $T_c = 52$ K prepared by the powder-in-tube method. Supercond Sci Technol 21:112001
66. Qi Y, Zhang X, Gao Z, Zhang Z, Wang L, Wang D, Ma Y (2009) Superconductivity of powder-in-tube $Sr_{0.6}K_{0.4}Fe_2As_2$ wires. Physica C 469:717
67. Zhang X, Wang L, Qi Y, Wang D, Gao Z, Zhang Z, Ma Y (2010) Effect of sheath materials on the microstructure and superconducting properites of $SmO_{0.7}F_{0.3}FeAs$ wires. Physica C 470:104
68. Wang L, Qi Y, Wang D, Zhang X, Gao Z, Zhang Z, Ma Y, Awaji S, Nishijima G, Watanabe K (2010a) Large transport critical currents of powder-in-tube $Sr_{0.6}K_{0.4}Fe_2As_2$/Ag superconducting wires and tapes. Physica C 470:183
69. Wang L, Qi Y, Wang D, Gao Z, Zhang X, Zhang Z, Wang C, Ma Y (2010b) Low-temperature synthesis of $SmFeAsO_{0.7}F_{0.3-\delta}$ wires with a high transport critical current density. Supercond Sci Technol 23:075005
70. Wang L, Qi Y, Zhang Z, Wang D, Zhang X, Gao Z, Yao C, Ma Y (2010c) Influence of Pb addition on the superconducting properties of polycrystalline $Sr_{0.6}K_{0.4}Fe_2As_2$. Supercond Sci Technol 23:054010
71. Ma Y, Wang L, Qi Y, Gao Z, Wang D, Zhang X (2011) Development of Powder-in-tube processed iron pnictide wires and tapes. IEEE Trans Appl Supercond 21:2878
72. Qi Y, Wang L, Wang D, Zhang Z, Gao Z, Zhang X, Ma Y (2010) Transport critical currents in the iron pnictide superconducting wires prepared by the ex situ PIT method. Supercond Sci Technol 23:055009
73. Togano K, Matsumoto A, Kumakura H (2011) Large transprot critical current densities of Ag sheathed (Ba, K)Fe_2As_2 + Ag superconducting wires fabricated by an ex-situ powder-in-tube process. Appl Phys Express 4:043101
74. Ding Q, Prombood T, Tsuchiya Y, Nakajima Y, Tamegai T (2012) Superconducting properties and magneto-optical imaging of $Ba_{0.6}K_{0.4}Fe_2As_2$ PIT wires with Ag addition. Supercond Sci Technol 25:035019
75. Fujioka M, Kota T, Matoba M, Ozaki T, Takano Y, Kumakura H, Kamihara Y (2011) Effective ex-situ fabrication of F-doped SmFeAsO wire for high transport critical current density. Appl Phys Express 4:063102
76. Zhang Q, Wang C, Yao C, Lin H, Zhang X, Wang D, Ma Y, Awaji S, Watanabe K (2013) Combined effect of Sn addition and post-rolling sintering on the superconducting properties of $SmFeAsO_{1-x}F_x$ tapes fabricated by an ex-situ powder-in-tube process. J Appl Phys 113:123902
77. Wang C, Yao C, Lin H, Zhang X, Zhang Q, Wang D, Ma Y, Awaji S, Watanabe K, Tsuchiya Y, Sun Y, Tamegai T (2013) Large transport J_c in Sn-added $SmFeAsO_{1-x}F_x$ tapes prepared by an ex situ PIT method. Supercond Sci Technol 26:075017

78. Wang L, Qi Y, Gao Z, Wang D, Zhang X, Ma Y (2010) The role of silver addition on the structural and superconducting properties of polycrystalline $Sr_{0.6}K_{0.4}Fe_2As_2$. Supercond Sci Technol 23:025027
79. Zhang X, Wang Q, Li K, Cai Y, Jiang F, Wang Z, Li J, Yao C, Lin H et al (2015) Investigation of J_c-suppressing factors in flat-rolled $Sr_{0.6}K_{0.4}Fe_2As_2$/Fe Tapes. IEEE Trans Appl Supercond 25:7300105
80. Gao Z, Ma Y, Yao C, Zhang X, Wang C, Wang D, Awaji S, Watanabe K (2012) High critical current density and low anisotropy in textured $Sr_{1-x}K_xFe_2As_2$ tapes for high field applications. Sci Rep 2:998
81. Wang L, Qi Y, Zhang X, Wang D, Gao Z, Wang C, Yao C, Ma Y (2011) Textured $Sr_{1-x}K_xFe_2As_2$ superconducting tapes with high critical current density. Physica C 471:1689
82. Gao Z, Wang L, Yao C, Qi Y, Wang C, Zhang X, Wang D, Wang C, Ma Y (2011) High transport critical current densities in textured Fe-sheathed $Sr_{1-x}K_xFe_2As_2$ superconducting tapes. Appl Phys Lett 99:242506
83. Weiss JD, Tarantini C, Jiang J, Kametani F, Polyanskii AA, Larbalestier DC, Hellstrom EE (2012) High intergrain critical current density in fine-grain $(Ba_{0.6}K_{0.4})Fe_2As_2$ wires and bulks. Nat Mater 11:682

Chapter 2
Fabrication and Characterization Methods

2.1 Basic Fabrication Process of Wires and Tapes

The iron-based superconducting wires and tapes in this thesis are prepared with a powder-in-tube (PIT) process, as illustrated in Fig. 2.1. A typical PIT procedure generally includes three steps: (1) packing the starting powders into metal tubes. For the in-situ PIT process, the starting powders are high-purity raw elements or compounds, and for the ex-situ PIT process, the starting powder is the reacted powder made by solid state reaction. (2) mechanical deformations such as drawing and rolling that make the packed metal tube into wires and tapes with a certain dimension. (3) a heat treatment process for eliminating the cold deformation induced stress and improving the grain connectivity of the superconducting phase.

The key facilities for the fabrication of iron-based superconducting wires and tapes are shown in Fig. 2.2. The preparation and weighing for raw materials such as Sr, Ba, K, Fe and As were done in a glove box, which was filled with high-purity Ar gas and can keep the raw materials from being oxidized. The raw materials were put in sealed jars with Ar gas inside, and then mixed by using a planetary ball milling. For the in-situ PIT process, the mixed raw materials were packed into metal tubes in the glove box. For the ex-situ PIT process, the mixed raw materials were packed into Nb tubes, sealed in Fe tubes by arc welding, and sintered in a furnace for solid-state reaction synthesis. The obtained precursors were ground into fine powders in the glove box, and then packed into metal tubes. Before the drawing process, tube composite can be pre-worked by rotary swaging to make it thinner and longer. After cold drawing with a reducing rate of about 10%, we can get round wires with a diameter less than 2 mm. The as-drawn wire can be further deformed into tapes with a thickness down to ~0.2 mm by using a flat rolling machine. The as-drawn wires or as-rolled tapes were cut into short samples of several centimeters in length, and then the samples were sealed in quartz tubes for heat treatment. The parameters of the heat treatment process such as temperature and duration were different for in-situ and ex-situ samples.

C. Yao, *Fabrication and Properties of High-Performance 122-Type Iron-Based Superconducting Wires and Tapes*, Springer Theses,
https://doi.org/10.1007/978-981-10-5184-5_2

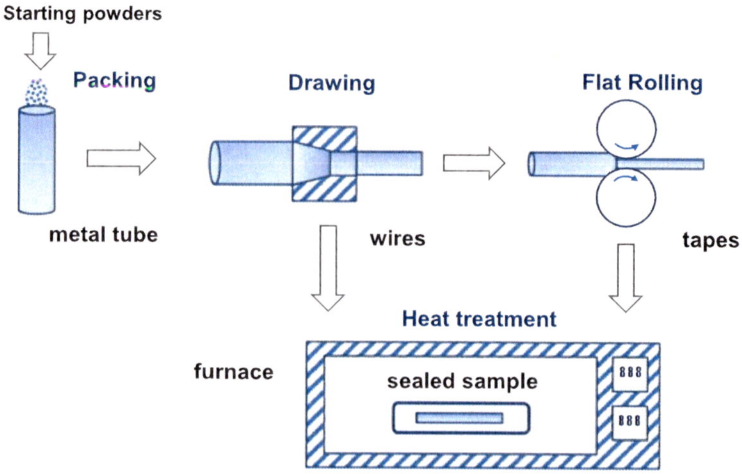

Fig. 2.1 Typical powder-in-tube (PIT) process for iron-based superconducting wires and tapes

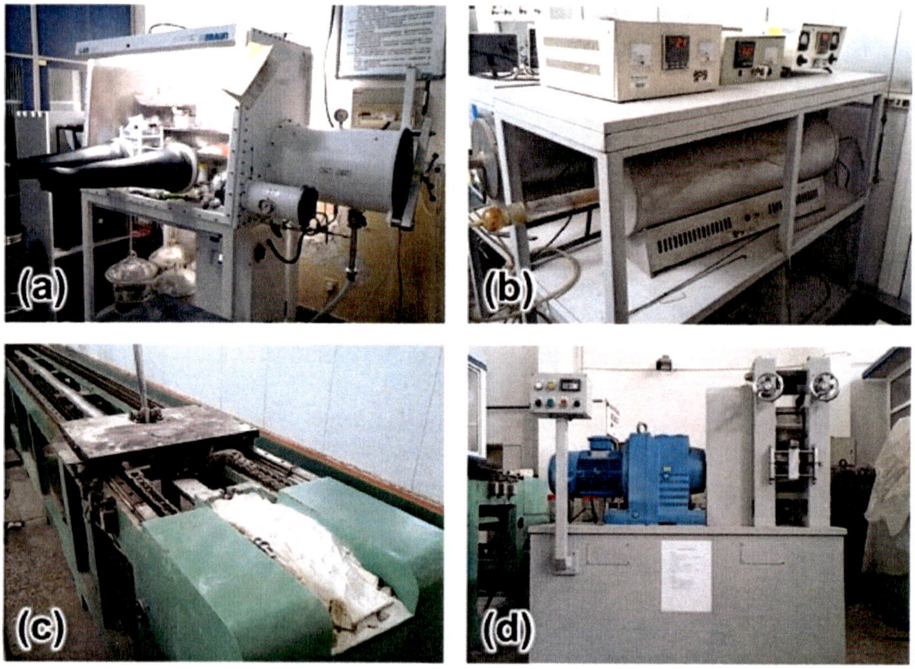

Fig. 2.2 Facilities for the fabrication of iron-based superconducting wires and tapes including, **a** glove box, **b** furnace, **c** drawing machine and **d** flat rolling machine

2.2 Characterization of Wires and Tapes

(1) Phase identification

The phases of the superconducting cores of wires and tapes were identified by powder x-ray diffraction (XRD) using a Rigaku D/MAX 2500 diffractometer with Cu Kα radiation. The superconducting cores were removed from wires and tapes by peeling off metal sheaths and then ground into fine powder in an agate mortar in a glove box. The obtained powders were put in the powder sample holder of diffractometer for XRD examination. The peaks in XRD patterns were indexed, and then the main phase and impurity phases can be identified.

(2) Microsturcture

The microstructure of wires and tapes was observed with an Olympus BX51 optical microscope, Zeiss SIGMA or Hitachi S4800 scanning electron microscopes (SEM), and a FEI Tecnai F20 high-resolution transmission electron microscope (HRTEM). For the observation of the cross sections of wires and tapes, the cut wire or tape samples were cold mounted with resin. The samples were then ground by abrasive papers and finally polished by polishing cloths with diamond paste. For the microstructure of grains and particles in superconducting cores, the superconducting cores were removed from wires and tapes by peeling off metal sheaths and then carefully broke into pieces. The cross sections of obtained pieces were submitted to SEM examination for material defects such as pores, cracks and impurities. In order to obtain the microstructural information of grain boundaries, the superconducting cores were polished and processed with focused ion beam (FIB) for a thickness small enough (~50 nm) for HRTEM examination at atomic scale, with which the secondary phased at grain boundaries and the grain boundary angles can be studied.

(3) Elemental composition

The elemental composition of materials can be investigated with an energy dispersive X-ray spectroscope (EDX) in conjunction with SEM or TEM. Except for determine the local elemental composition in a certain position, an elemental line scan or mapping function can also be used to reveal the change of composition or the inhomogeneity of composition, respectively.

(4) Microhardness

Since it is quite hard to measure the precise weight and volume to determine the mass density for the small superconducting core inside wires, microhardness was widely used as an index of the mass density. The cut wire or tape samples were cold mounted with resin, and then the samples were ground by abrasive papers and polished by polishing cloths with diamond paste. Vickers hardness was measured on the polished cross section of samples with 0.05 kg load and 10 s duration using a Wilson Vickers 402 MVD microhardness tester.

(5) Grain texture

For the evaluation of the degree of c-axis texture for superconducting phase, the superconducting cores were removed from tapes by peeling off metal sheaths, carefully ground and polished by abrasive papers and lapping films, and finally submitted for XRD examinations. The degree of texture can be quantified using a parameter F calculated from the XRD patterns obtained on the polished surface (parallel to the rolling plane) by the Lotgering method [1]:

$$F = \frac{\rho - \rho_0}{1 - \rho_0} \tag{2.1}$$

where $\rho = \Sigma I(00\ l)/\Sigma I(hkl)$, $\rho_0 = \Sigma I_0(00\ l)/\Sigma I_0\ (hkl)$, I and I_0 are the intensities of each reflection peak (hkl) for the textured and randomly oriented powder samples, respectively.

(6) Superconducting transition temperature

The superconducting transition temperature (T_c) of samples can be determined through the temperature dependence of resistance (R-T) or magnetization (M-T). The resistance measurements were carried out on a Quantum Design physical property measurement system (PPMS), as shown in Fig. 2.3, using a four-probe method. The $T_{c,\ \mathrm{onset}}$ and $T_{c,\ \mathrm{zero}}$ are determined by the intersection of straight lines extrapolated from the transition region with the normal state and superconducting state on R-T curves, respectively. The M-T measurements were performed in a vibrating sample magnetometer (VSM) in conjunction with PPMS. The sample is first cooled below the $T_{c,\ \mathrm{zero}}$ in the absence of magnetic fields, then the magnetization is measured

Fig. 2.3 Photograph of physical property measurement system with a vibrating sample magnetometer

while increasing the temperature in an applied field of 20 Oe (i.e. zero filed cooling: ZFC). After the temperature is above the $T_{c,\text{ onset}}$, the magnetization is measured again while decreasing the temperature below the $T_{c,\text{ zero}}$ (i.e. field cooling: FC). Due to the percolation of superconductivity, the diamagnetic transition always emerges at temperatures slightly lower than $T_{c,\text{ onset}}$ obtained by the *R*-*T* measurement.

(7) Upper critical field and irreversibility field

The *R*-*T* measurement is carried out on PPMS at various magnetic fields. The upper critical field H_{c2} and irreversibility field H_{irr} are estimated with the criteria of 90% and 10% of resistivity at normal state respectively. The upper critical field at zero-temperature $H_{c2}(0)$ was calculated using the Werthamer-Helfand-Hohenberg (WHH) formula [2]:

$$H_{c2}(0) = -0.693T_c\frac{dH_{c2}}{dT} \tag{2.2}$$

(8) Magnetic critical current density

The magnetic hysteresis measurements were carried out in a vibrating sample magnetometer (VSM) in conjunction with PPMS. The bulk magnetic J_c can be derived from the magnetic hysteresis loops using the Bean model [3]:

$$J_c = \frac{20\Delta M}{a\left(1-\frac{b}{3a}\right)} \tag{2.3}$$

where ΔM is the difference between the magnetization when sweeping magnetic fields up and down, a and b are the sample widths ($a < b$).

(9) Transport critical current density

The transport critical current density (J_c) is defined as the quotient of critical current (I_c) divided by the area of superconducting cores measured on the transverse cross section of superconducting wires and tapes. In this thesis, the test of transport I_c under various external magnetic fields at 4.2 K was performed in a 15 T superconducting magnet in High Field Laboratory for Superconducting Materials, Institute for Materials Research, Tohoku University under the direction of Prof. Awaji by using the standard four-probe method. The transport I_c measurement system is presented in Fig. 2.4. The length of the wire and tape samples is about 3 cm, and the criterion used for evaluating the critical point is 1 μV/cm.

(10) Magneto-optical imaging

The magneto-optical imaging (MOI) characterization was conducted in the Prof. Tamegai group at the University of Tokyo. The MOI system is shown in Fig. 2.5. After removing the superconducting cores of tapes from by peeling off metal sheaths, and polishing them with lapping films, the Sr-122 samples

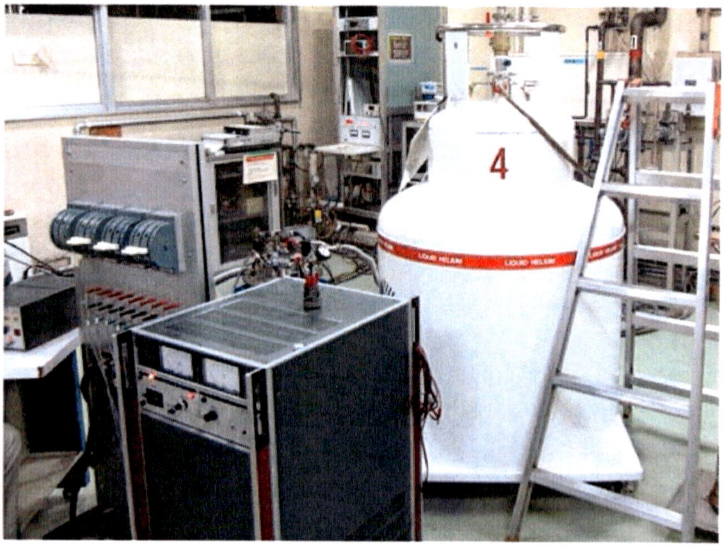

Fig. 2.4 Photograph of the transport I_c measurement system in Institute for Materials Research, Tohoku University

Fig. 2.5 Photograph of the magneto-optical imaging system in the University of Tokyo

were cut into small rectangular pieces with dimensions of about 0.82 × 0.80 × 0.10 mm^3. A Bi-substituted iron-garnet indicator film was placed in direct contact with the sample, and the whole assembly was attached to the cold finger of a He-flow cryostat (Microstat-HR, Oxford Instruments). The MO images of the reflected light were obtained by using a cooled CCD camera with 12-bit resolution (ORCA-ER, Hamamatsu). A differential imaging technique was used to enhance the visibility of the local magnetic induction and remove the contrasts induced by the impurity phases and scratches of the garnet film.

References

1. Lotgering FK (1959) Topotactical reactions with ferrimagnetic oxides having hexagonal crystal structures-I. J Inorg Nucl Chem 9:113
2. Werthamer NR, Helfand E, Hohenberg PC (1966) Temperature and purity dependence of the superconducting critical field, H_{c2}. III. Electron spin and spin-orbit effects. Phys Rev 147:295–302
3. Bean CP (1964) Magnetization of high-field superconductors. Rev Mod Phys 36:31

Chapter 3
Effect of Metal Additives on Ba-122 Tapes

The transport critical current density (J_c) of iron-based superconducting wires and tapes prepared with the powder-in-tube (PIT) process was seriously limited by extrinsic material defects such as impurity phases, micro cracks, loose microstructure and composition segregation. These defects lead to low grain connectivity and superconducting properties, thus resulting in a lowered transport J_c. In order to improve the J_c performance, the fabricating process of iron-based superconducting (IBS) wires and tapes was being optimized. For example, ex-situ PIT process was used instead of the previous in-situ process, the heat treatment condition was improved, and metal additives were employed.

For 1111-type iron-based superconducting wires such as LaFeAsO and SmFeAsO wires, the heat treatment temperature was 900–1180 °C, and the time for heat treatment was up to 30–48 h [1–6]. For 122-type iron-based superconducting wires such as $Sr_{1-x}K_xFe_2As_2$ (Sr-122) wires, the heat treatment temperature was 850–900 °C, and the time for heat treatment was also longer than 30 h [7–9]. After the ex-situ PIT process was proposed, the grain connectivity and composition homogeneity was remarkably improved, but the time of the final heat treatment for wires and tapes was still up to 20–30 h [10, 11]. During such long-time annealing processes, the superconducting phase will lost some volatile elements such as potassium, fluorine and arsenic, which can cause composition segregation and pores. In 2011, a short-time high-temperature annealing technique was introduced for the heat treatment for Sr-122 tapes. This new process has a much higher temperature of 1100 °C, which is higher than the melting point of FeAs phase (~1050 °C), leading to better grain coupling. Due to the high annealing temperature, the time of heat treatment was significantly shortened as 5–15 min, resulting in smaller grains and much fewer residual pores. As a result, the transport J_c of Sr-122 tapes processed with the short-time high-temperature annealing technique was increased to 5.4×10^3 A/cm^2 at 4.2 K in self field [12]. On the other hand, there are still some pores and gaps inside the superconducting phase, for which the grain connectivity needs to be improved.

C. Yao, *Fabrication and Properties of High-Performance 122-Type Iron-Based Superconducting Wires and Tapes*, Springer Theses,
https://doi.org/10.1007/978-981-10-5184-5_3

It was found that metal additives such as silver and lead can enhance the J_c of Sr-122 polycrystalline bulks and wires [8–10, 12]. By investigating the superconducting properties and microstructure, we found that due to the low melting point of 327.5 °C, the added lead will be in liquid state during the heat treatment process, so it can bridge the pores and cracks to enhance grain connection. The lead addition can increase the transport J_c in low-field region, but hardly improve the J_c performance in high fields. Silver addition can not only prevent the formation of glassy phases and amorphous layers at grain boundaries, but also promote the crystallization of superconducting phase. Therefore, the transport J_c at high-field region can be enhanced by silver addition [8, 10, 13].

As introduced in Chap. 1, with similar lattice structure and high transition temperatures to Sr-122, $Ba_{1-x}K_xFe_2As_2$ (Ba-122) is also another promising candidate for practical applications among 122-type iron-based superconductors. Therefore, it is of significance to investigate the effect of metal addition on Ba-122 tapes. Here in this thesis, the co-addition effect of silver and lead was systematically investigated in comparison with that of silver or lead single addition for Ba-122 IBS tapes. It is expected that by combining the advantages of these two metal additives, transport J_c can be improved in the whole field region. Furthermore, the effect of annealing time of the short-time high-temperature annealing technique on microstructure and superconducting properties was studied for the Ba-122 tapes with single- and co-addition.

3.1 Composition and Phase Analysis

The $Ba_{1-x}K_xFe_2As_2$ precursors were prepared with small Ba pieces, K bulks and fine FeAs powder according to a stoichiometric ratio of 0.66:0.48:2.00. These starting materials were mixed and then put in an Al_2O_3 crucible, which was immediately sealed in a stainless steel pipe by argon arc welding. Then the pipe was put into a tubular furnace maintained at a temperature at 1100 °C for 10 min and cooled to room temperature by switching off the furnace. The precursor was ground into powder in an agate mortar and packed in an Nb tube, which was sealed in a stainless steel pipe and sintered at 900 °C for 20 h. Then the obtained product was ground into fine powder again and added with silver and lead powders following the stoichiometry of $Ba_{0.66}K_{0.48}Fe_2As_2 + Ag_{0.5}$, $Ba_{0.66}K_{0.48}Fe_2As_2 + Pb_{0.2}$ and $Ba_{0.66}K_{0.48}Fe_2As_2 + Ag_{0.5} + Pb_{0.2}$, respectively. Finally, the mixture was packed in iron tubes of 8 mm in outer and 5 mm in inner diameter. These billets were sequentially rotary swaged, drawn to wires, and rolled into 0.7 mm thick tapes. The tapes were cut into short samples of 6 cm in length, sealed into iron tubes and then annealed at 1100 °C for 5, 10 and 15 min, respectively. The samples are listed in Table 3.1.

Because of the significantly shortened annealing time, a clear boundary between the superconducting core and the iron sheath can be observed, indicating that there is no obvious reaction at the inner surface of the sheath during the heat treatment, as shown in Fig. 3.1. To further investigate the element distribution, an EDX line-scan

Table 3.1 Information of Ba-122 tapes with metal additions

Stoichiometry	Sample name	Annealing time (min)	$T_{c,\ onset}$ (K)	Transport J_c (kA/cm^2) (4.2 K 0 T)
$Ba_{0.66}K_{0.48}Fe_2As_2 + Ag_{0.5}$	Ag-5	5	33.8	0.21
	Ag-10	10	33.8	0.23
	Ag-15	15	33.9	0.27
$Ba_{0.66}K_{0.48}Fe_2As_2 + Pb_{0.2}$	Pb-5	5	31.4	6.43
	Pb-10	10	31.1	4.29
	Pb-15	15	31.2	4.71
$Ba_{0.66}K_{0.48}Fe_2As_2 + Ag_{0.5} + Pb_{0.2}$	AgPb-5	5	33.7	14.29
	AgPb-10	10	34.0	11.86
	AgPb-15	15	33.9	8.43

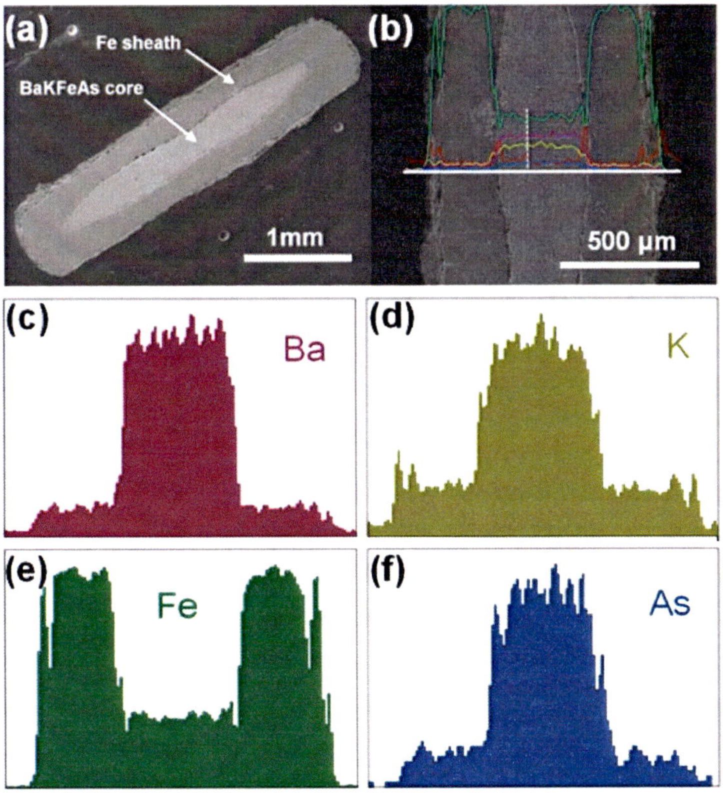

Fig. 3.1 **a** SEM image of the cross section of Ba-122 tape without metal additives. The superconducting core and the Fe sheath are indicated by arrows. **b** EDX line-scan on the cross section of the tape. The scanning path is indicated by the white solid line. From top to bottom, the color curves crossing the white dashed line represent intensities of Fe, Ba, K, O and As elements respectively. Reprinted from Yao et al. [14]. The enlarged EDX pattern for each element is shown for **c** Ba, **d** K, **e** Fe and **f** As

was performed on the cross section of a pure Ba-122 tape. It can be seen that the Ba, K and As elements are successfully restricted within iron sheaths, and for each element, the content has a dramatic change at the boundary of the superconducting core.

Figure 3.2 shows the EDX element mapping for Ba-122 tapes with silver and lead co-addition heat treated at 1100 °C for 5 min. It is clear that all the elements did not diffuse into iron sheaths, which is in accordance with the results by EDX line-scan. Moreover, Ba, K, Fe and As elements are homogeneously dispersed in superconducting cores, indicating no obvious composition segregation for the main Ba-122 phase. In contrast, the distribution of added Ag and Pb elements is very inhomogeneous. It can be concluded that the Ag and Pb elements did not enter into Ba-122 grains, and they just stay at the grain boundaries and in some pores and gaps between Ba-122 grains.

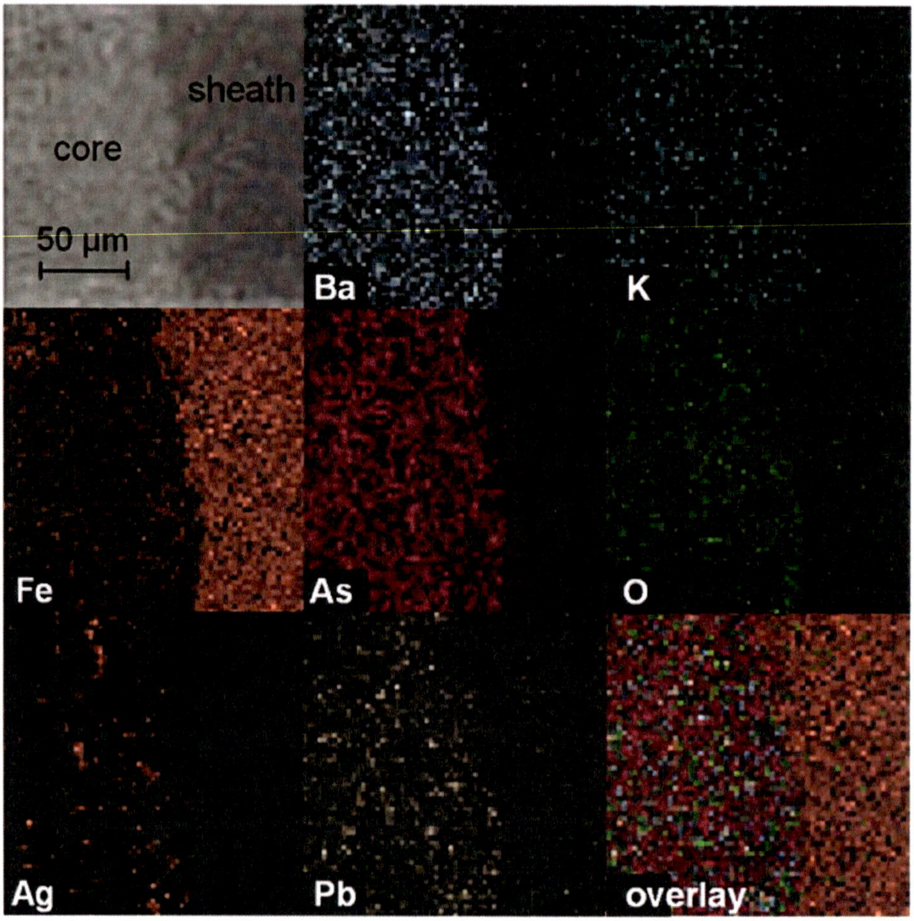

Fig. 3.2 EDX mapping images on the cross section of Ba-122 tapes with Ag and Pb co-addition. Reproduced from Yao et al. [14]

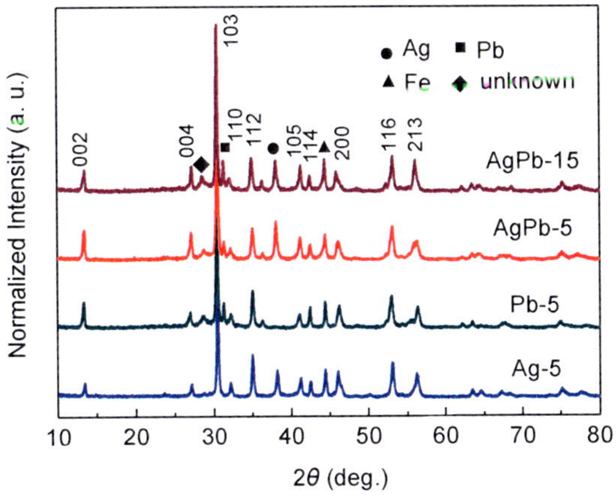

Fig. 3.3 XRD patterns of samples measured after grounding the superconducting core into powders. The main peaks of the Ba-122 phase are indexed, and the peaks of impurity phases are marked with different symbols. Reproduced from Yao et al. [14]

The powder XRD patterns of several samples are presented in Fig. 3.3. Except for some minor impurity peaks, the diffraction patterns show strong peaks of Ba-122 phase, indicating it is the main phase for all the samples. The Ag or Pb peak can be respectively detected in the tapes with single addition, and both of them were simultaneously observed in the XRD pattern of the samples with co-addition. The Fe impurity phase detected in all the samples was introduced by the iron sheaths of the tapes. It is noticed that the XRD patterns of BaKFeAs–AgPb-5 and BaKFeAs–AgPb-15 are almost the same, suggesting that the time of high-temperature annealing has no significant effect on the phases of the samples. The observed Ag and Pb peaks, and the result that the peak positions in the XRD patterns are exactly the same for samples with different metal additions both indicate that the Ag and Pb additives did not go into the Ba-122 lattice, which is in accordance with the analysis of EDX element mapping.

3.2 Superconducting Properties

Figure 3.4 gives the temperature dependence of the resistivity normalized by resistivity at 300 K for samples with various additives annealed at 1100 °C for 5 min. The resistance measurements were carried out based on the Ba-122 cores after removing the iron sheaths using a four-probe method. The onset superconducting critical temperature T_c of BaKFeAs–Ag-5, BaKFeAs–Pb-5 and BaKFeAs–AgPb-5 tape is 33.8 K, 31.4 K and 33.7 K, respectively. It can be observed that the the $T_c s$ of samples with Ag additive are higher than the that of other ones, and the same conclusion can be derived from the T_c data of samples annealed for 10 and 15 min, indicating that Ag additive has a positive effect on the T_c of samples. This is similar to the effect of Ag addition on $YBa_2Cu_3O_x$ (YBCO) superconductors, in which Ag can promote the formation of superconducting phase and the growth of grains [15, 16]. On the other hand, as shown in Fig. 3.5, we can see that the supercon-

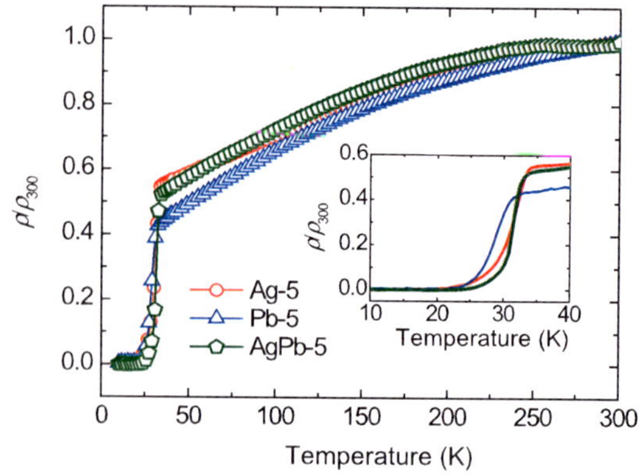

Fig. 3.4 Temperature dependence of resistivity normalized by resistivity at 300 K for the Ba-122 tapes with various additives annealed for 5 min. The inset gives an expanded view at low temperature

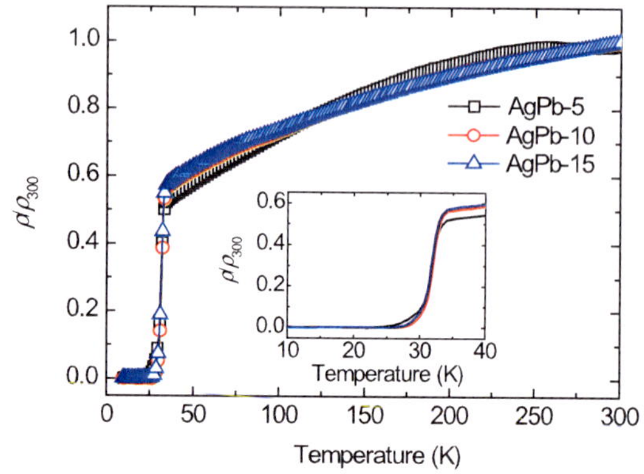

Fig. 3.5 Temperature dependence of resistivity normalized by resistivity at 300 K for the Ba-122 tapes with Ag and Pb co-addition annealed for 5–15 min. The inset gives an expanded view at low temperature

ducting transition of BaKFeAs–AgPb-10 and BaKFeAs–AgPb-15 is a little steeper than BaKFeAs–AgPb-5. This indicates that a longer annealing time can improve the crystallinity of the superconducting phase, but its improvement on T_c seems smaller than that caused by Ag addition.

In order to further understand the co-addition effect of Ag and Pb on superconducting properties, the temperature dependent resistivity of BaKFeAs–Pb-5 and BaKFeAs–AgPb-5 samples in various magnetic fields ($B = 0$, 1, 3, 5, 7 and 9 T) was studied. The applied fields in were parallel to the sample plane. As presented in Fig. 3.6, with the increase of magnetic fields, the superconducting transition curves of these two samples shifts towards lower temperature region, and become less steep. Their upper critical field H_{c2} and irreversibility field H_{irr}, which are estimated with the criteria of 90% and 10% of resistivity at normal state respectively, are shown in Fig. 3.7. The upper critical field at zero-temperature $H_{c2}(0)$ was calculated by using the Werthamer-Helfand-Hohenberg (WHH) formula (see Chap. 2 for detail). For BaKFeAs–Pb-5 and BaKFeAs–AgPb-5 samples, taking $T_c = 28.5$ K and 31.5 K, the $H_{c2}(0)$ is 124 T and 129 T respectively. It is obvious that the H_{c2} and H_{irr} of

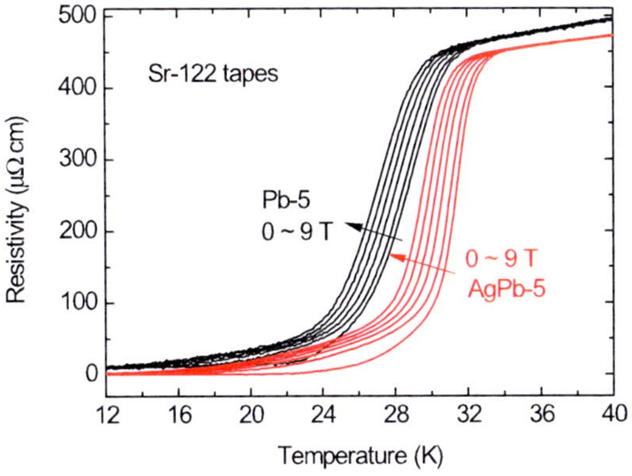

Fig. 3.6 Temperature dependence of resistivity at various magnetic fields from 0 to 9 T for Ba-122 tapes with Pb addition and Pb-Ag co-addition annealed for 5 min. The applied field is parallel to the tape plane

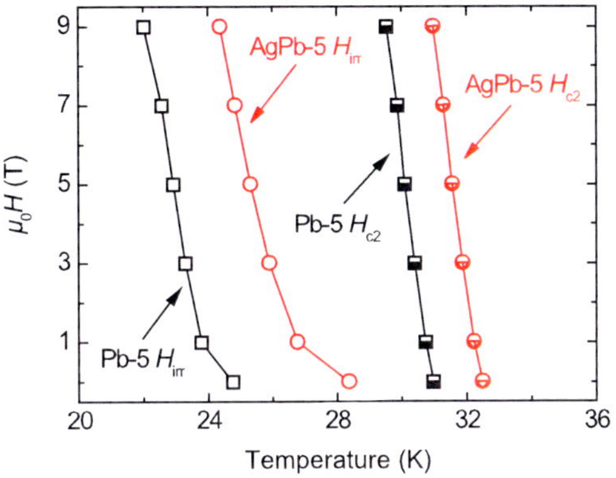

Fig. 3.7 Temperature dependence of the upper critical field H_{c2} and irreversibility field H_{irr} for for the Ba-122 tapes with Pb addition and Pb–Ag co-addition annealed for 5 min. Reproduced from Yao et al. [14]

BaKFeAs–AgPb-5 is much higher than that of BaKFeAs–Pb-5 at the same temperature, indicating that compared with the Pb addition, the Ag and Pb co-addition can further improve the superconducting properties of Ba-122 tapes.

The transport I_c was measured at 4.2 K using short tape samples of 3 cm in length with standard four-probe method and evaluated by the criterion of 1 μV/cm. The applied fields in transport I_c measurement are parallel to the tape surface. The transport critical I_c of tapes with various metal additives and annealing times in self field at 4.2 K is summarized in Fig. 3.8. In general, the I_c decreases with the increase of annealing time. For the samples heat-treated for the same times, the I_c value in self field can be significantly raised by Pb addition. By Ag and Pn co-addition, the I_c value can be approximately doubled compared with that by Pb addition. The I_c of tape samples with Ag and Pb co-addition annealed for 5 min reached 100 A, and correspondingly the J_c was above 1.4×10^4 A/cm^2 at 4.2 K and in self field. Since the cross sectional areas of the superconducting cores for tapes with various additives and heat treatment conditions are almost the same, the difference between J_c values is similar to that between I_c values.

As presented in Table 3.1, the J_c values of the tapes with Ag addition are above 200 A/cm^2 at 4.2 K and in self field, which are on the same level (10^2–10^3 A/cm^2) of that

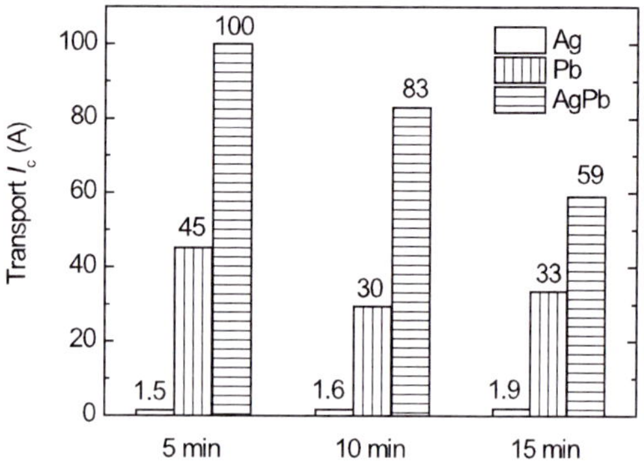

Fig. 3.8 Transport I_c at 4.2 K in self-field for Ba-122 tapes with Ag addition, Pb addition and Pb-Ag co-addition annealed for 5, 10 and 15 min, respectively

of previously reported Sr-122 tapes with Ag addition [8], though in that work the tapes were annealed at 850–900 °C for 35 h, and the amount of Ag addition varied between 5 and 20 wt.%. The field dependent J_c at 4.2 K of Pb doped and Ag–Pb co-doped tapes is shown in Fig. 3.9. It is obvious that the transport J_c for all the samples drop dramatically in low field region, by about one order of magnitude. This is similar to the J_c of cobalt-doped $BaFe_2As_2$ bicrystal films with large misorientation angles [17], indicating a characteristic of weak link in these tape samples. However, the decrease of J_c become much slower for the samples with Ag and Pb co-addition compared with that of the samples with Ag addition when the applied field is increased above 1 T. Therefore, Ag and Pb co-addition can not only enhance the J_c performance in self field, but also help to weaken the field dependence of in-field J_c. The J_c value of BaKFeAs-AgPb-5 sample is still above 200 A/cm^2 at 4.2 K and 6 T. The inset of Fig. 3.9 shows the J_c curves of this sample measured in increasing and decreasing fields. The hysteretic effect of J_c is relevant to penetration of flux into strong pinning intragranular regions, and the presence of intragranular I_c can enhance intergranular I_c when the field is decreasing [18], indicating that there are still weak-linked current paths between grains in these tapes.

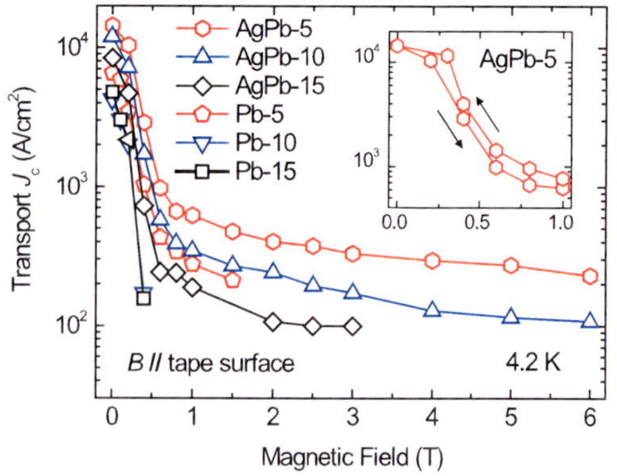

Fig. 3.9 Field dependence of the transport J_c of Ba-122 tapes with Pb addition and Ag-Pb co-addition at 4.2 K. The inset shows the J_c of BaKFeAs–AgPb-5 sample successively measured in increasing and decreasing fields. Reprinted from Yao et al. [14]

3.3 Microstructure of Superconducting Phase

To investigate the effect of different metal additives and annealing time on the microstructure of Ba-122 tapes after the short-time high-temperature annealing, the crushed superconducting cores were analyzed with a SEM, as shown in Fig. 3.10. In Fig. 3.10a, there are some large Ag particles (marked by arrows) embedding in the Ba-122 matrix in the BaKFeAs–Ag-5 sample. These inhomogeneously dispersed Ag particles may be caused by the melting and segregation of Ag during the heat treatment process. In Fig. 3.10b, it can be seen that the Pb additives (marked by arrows) bind the Ba-122 phase tightly in the BaKFeAs–Pb-5 sample, but their distribution in Ba-122 matrix is also not very homogeneous. The inhomogeneous distribution of Ag and Pb additives is in accordance with the result of EDX examination. The average size of the Ba-122 particles in the BaKFeAs–Ag-5 sample is about 1–2 μm, smaller than the ~4 μm for the BaKFeAs–Pb-5 sample. Smaller particle size means a high density of grain boundaries, which usually cause weak grain connectivity and a low transport J_c, especially in low field region. However, in Fig. 3.10a, it is clear that the crystallization of Ba-122 phase for the BaKFeAs-Ag-5 sample is better than that for the BaKFeAs–Pb-5 sample in Fig. 3.10b. The improvement on crystallization is

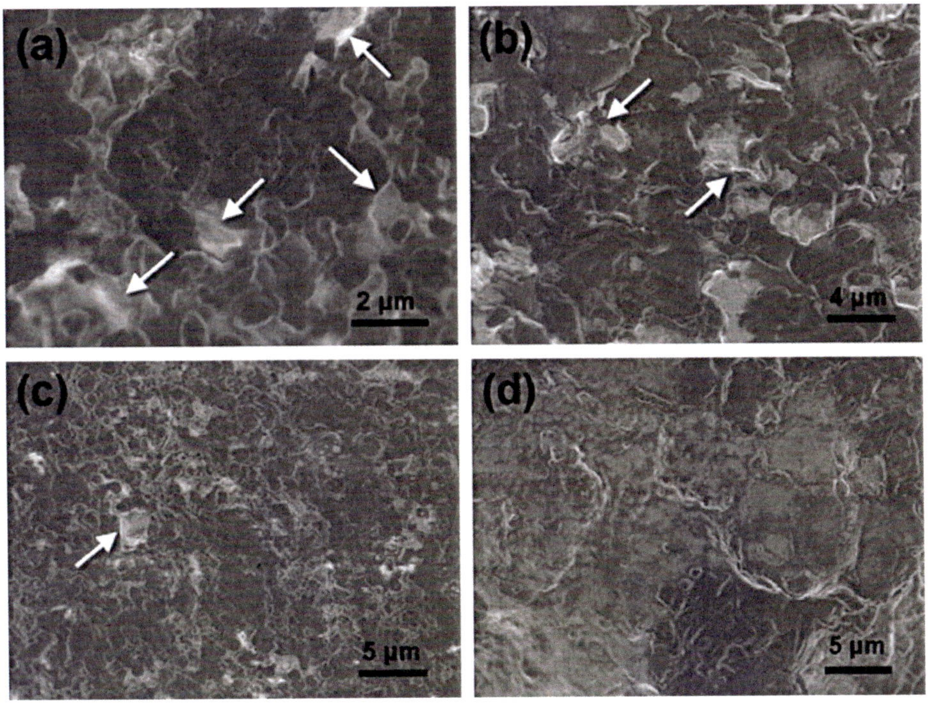

Fig. 3.10 SEM images for the superconducting cores of **a** BaKFeAs–Ag-5, **b** BaKFeAs–Pb-5, **c** BaKFeAs–AgPb-5 and **d** BaKFeAs–AgPb-15 tapes. The metallic additive segregations are marked by arrows. Reprinted from Yao et al. [14]

due to the reduced glassy phases and amorphous layers at grain boundaries by Ag addition [8, 13]. A better crystallization can help to enhance the in-field performance of current carrying capability.

For the BaKFeAs–AgPb-5 sample, it can also be found some large metallic particles formed by Ag and/or Pb additives, as marked by arrows in Fig. 3.10c. The size of Ba-122 particles is 1–3 μm for this sample, with a similar crystallization to the BaKFeAs-Ag-5 sample, showing a microstructure with combined features of Ag and Pb single additions. Accordingly, for the samples with Ag and Pb co-addition, it can be concluded that Pb doping can improve the particle connections, while Ag doping is helpful for grain formation. As a result, taking the advantages of these two metal additives, the transport J_c of the BaKFeAs–AgPb-5 sample can be enhanced in the whole field region. When comparing the BaKFeAs–AgPb-5 and BaKFeAs–AgPb-15 samples shown in Fig. 3.10c, d, we can see that with the increase of annealing time, the small Ba-122 particles grew into large irregular bulks, which may lead to a relatively worse crystallinity. Moreover, the metallic additive phase in Fig. 3.10d become much less, which may be caused by the element evaporation during a longer heat treatment process, and will weaken the positive effect on improving grain connectivity by metal additives. Therefore, the transport J_c of the BaKFeAs–AgPb-15 becomes lower than that of the BaKFeAs–AgPb-5 sample.

3.4 Conclusion

In conclusion, we have fabricated Fe-sheathed $Ba_xK_{1-x}Fe_2As_2$ superconducting tapes by an ex-situ powder-in-tube method combined with a short-time high-temperature annealing as the final heat treatment after the cold deformation. EDX mapping and line-scan indicated that there is no reaction layer between the Fe sheath and the superconducting core. With Ag and Pb co-addition, the transport J_c of Ba-122 tapes was effectively enhanced in the whole field region. Through SEM microstructural investigations, it can be concluded that Pb doping can improve the particle connections, while Ag doping is beneficial to the crystallization of Ba-122 phase. Compared to Ag and Pb single addition, the Ag and Pb co-addition successfully combined their own advantages, thus achieving a further improvement of current carrying capability. The annealing time for the heat treatment process was optimized, the best transport I_c reached 100 A at 4.2 K in self field for the sample annealed at 1100 °C for 5 min, and accordingly J_c value can be calculated as 1.4×10^4 A/cm^2 at 4.2 K in self field. The J_c curves of this sample measured in increasing and decreasing fields exhibit a hysteretic effect, indicating there are still weak-linked current paths between grains, hence there is still much room for the enhancement of J_c performance for 122-type iron-based superconducting wires and tapes.

References

1. Gao Z, Wang L,Qi Y, Wang D, Zhang X, Ma Y (2008) Preparation of $LaFeAsO_{0.9}F_{0.1}$ wires by the powder-in-tube method. Supercond Sci Technol 21:105024
2. Gao Z, Wang L, Qi Y, Wang D, Zhang X, Ma Y, Yang H, Wen H (2008) Superconducting properties of granular $SmFeAsO_{1-x}F_x$ wires with $T_c = 52$ K prepared by the powder-in-tube method. Supercond Sci Technol 21:112001
3. Wang L, Qi Y, Wang D, Gao Z, Zhang X, Zhang Z, Wang C, Ma Y (2010) Low-temperature synthesis of $SmFeAsO_{0.7}F_{0.3-\delta}$ wires with a high transport critical current density. Supercond Sci Technol 23:075005
4. Ma Y, Wang L, Qi Y, Gao Z, Wang D, Zhang X (2011) Development of powder-in-tube processed iron pnictide wires and tapes. IEEE Trans Appl Supercond 21:2878
5. Wang C, Yao C, Zhang X, Gao Z, Wang D, Wang C, Lin H, Ma Y, Awaji S, Watanabe K (2012) Effect of starting materials on the superconducting properties of $SmFeAsO_{1-x}F_x$ tapes. Supercond Sci Technol 25:035013
6. Chen YL, Cui YJ, Yang Y, Zhang Y, Wang L, Cheng CH, Sorrell C, Zhao Y (2008) Peak effect and superconducting properties of $SmFeAsO_{0.8}F_{0.2}$ wires. Supercond Sci Technol 21:115014
7. Qi Y, Zhang X, Gao Z, Zhang Z, Wang L, Wang D, Ma Y (2009) Superconductivity of powder-in-tube $Sr_{0.6}K_{0.4}Fe_2As_2$ wires. Physica C 469:717
8. Wang L, Qi Y, Wang D, Zhang X, Gao Z, Zhang Z, Ma Y, Awaji S, Nishijima G, Watanabe K (2010) Large transport critical currents of powder-in-tube $Sr_{0.6}K_{0.4}Fe_2As_2$/Ag superconducting wires and tapes. Physica C 470:183
9. Wang L, Qi Y, Zhang Z, Wang D, Zhang X, Gao Z, Yao C, Ma Y (2010) Influence of Pb addition on the superconducting properties of polycrystalline $Sr_{0.6}K_{0.4}Fe_2As_2$. Supercond Sci Technol 23:054010
10. Qi Y, Wang L, Wang D, Zhang Z, Gao Z, Zhang X, Ma Y (2010) Transport critical currents in the iron pnictide superconducting wires prepared by the ex situ PIT method. Supercond Sci Technol 23:055009
11. Togano K, Matsumoto A, Kumakura H (2011) Large transprot critical current densities of Ag sheathed (Ba, K)Fe_2As_2 + Ag superconducting wires fabricated by an ex-situ powder-in-tube process. Appl Phys Express 4:043101
12. Wang L, Qi Y, Zhang X, Wang D, Gao Z, Wang C, Yao C, Ma Y (2011) Textured $Sr_{1-x}K_xFe_2As_2$ superconducting tapes with high critical current density. Physica C 471:1689
13. Wang L, Qi Y, Gao Z, Wang D, Zhang X, Ma Y (2010) The role of silver addition on the structural and superconducting properties of polycrystalline $Sr_{0.6}K_{0.4}Fe_2As_2$. Supercond Sci Technol 23:025027
14. Yao C, Wang C, Zhang X, Wang L, Gao Z, Wang D, Wang C, Qi Y, Ma Y, Awaji S, Watanabe K (2012) Improved transport critical current in Ag and Pb co-doped $Ba_xK_{1-x}Fe_2As_2$ superconducting tapes. Supercond Sci Technol 25:035020
15. Lee DF, Chaud X, Salama K (1991) Transport current density in bulk oriented-grained $YBa_2Cu_3O_x$/silver composites. Phyisca C 181:81
16. Zhao Y, Cheng CH, Wang JS (2005) Roles of silver doping on joins and grain boundaries of melt-textured YBCO superconductor. Supercond Sci Technol 18:S34
17. Katase T, Ishimaru Y, Tsukamoto A, Hiramatsu H, Kamiya T, Tanabe K, Hosono H (2011) Advantageous grain boundaries in iron-pnictide superconductors. Nat Commun 2:409
18. Mchenry ME, Maley MP, Willis JO (1989) Systematics of transport critical-current-density hysteresis in polycrystalline Y-Ba–Cu–O. Phys Rev B 40:2666

Chapter 4
Mechanical Deformation and Grain Texture of Sr-122 Tapes

The material defects such as impurity phases, micro cracks and pores limit the transport currents by reducing the effective current paths. However, even for two tightly bound superconducting grains, because of the mismatched crystal orientation, there may be degradation for the transport currents passing across the grain boundary (GB) between them. For example, the transport J_c exponentially decreases at GBs with misorientation angles above 3–5° in YBCO superconductors, which is called weak-linked GBs. Such an exponential decay of J_c as a function of the misorientation angle was also observed in iron-based superconductors, but the critical GB angle for them is about 9°, larger than that of YBCO [1]. According to the experiments on $Ba_{1-x}Fe_{2-x}Co_xAs_2$ and $Ba_{1-x}Fe_2As_{2-x}P_x$ bicrystal films, even an external field of 10–10^3 Oe will significantly lower the inter-grain J_c pass through large-angle GBs [2–4]. As mentioned in Chap. 3, the hysteretic effect of the transport J_c measured in increasing and decreasing fields indicates the existence of weak-linked GBs in $Ba_{1-x}K_xFe_2As_2$ (Ba-122) tapes. Therefore, besides improving the grain connectivity, it is necessary to reduce the large-angle GBs to alleviate the weak-linking effect, so as to further enhance the J_c performance of iron-based superconducting tapes.

An effective solution for reducing the ratio of high-angle GBs is to make the grains textured. For high-J_c YBCO tapes, highly textured grains with a small in-plane misalignment were achieved by the so-called coated conductor technology. REBCO films are deposited on textured buffer layers, which are grown on metal substrates using an ion-beam-assisted deposition (IBAD) technique, or on rolling assisted biaxially textured metal substrates (RABiTS) [5–8]. A simpler route to induce grain texture is deform the powder-in-tube wire into tapes by flat rolling, which has been employed for the manufacture of powder-in-tube (PIT) Bi-2223 superconducting tapes [9–11]. Different from the bi-axial texture for coated conductors, the rolling induced texture is just along *c*-axis, but it has been proved to be quite successful for developing high-performance commercial Bi-2223 superconducting tapes. Since iron-based superconducting wires and tapes have been prepared by using

C. Yao, *Fabrication and Properties of High-Performance 122-Type Iron-Based Superconducting Wires and Tapes*, Springer Theses,
https://doi.org/10.1007/978-981-10-5184-5_4

the powder-in-tube method, and the crystalline structure of iron-based superconductors also has a two-dimensional character, their stratiform grains were expected to be textured by flat rolling process. The flat rolling induced grain texture was first observed in Fe-sheathed Sr-122 tapes with 0.8 mm in thickness, and the rapid drop of transport J_c at low field region was considerably alleviated [12].

Therefore, grain texture is an important means to improve the J_c performance of iron-based superconducting tapes. In this chapter, the grain texture of Sr-122 tapes with different mechanical deformation processes such as flat rolling and uniaxial press and different sheath materials including iron and silver are studied, a combined deformation processes of rolling and pressing is proposed, and the relationship between texture and J_c performance is discussed.

4.1 Fe-Sheathed Tapes Processed with Flat Rolling

Before start to investigate the mechanical deformation and grain texture of iron-based superconducting tapes, the precursors, metal additions and heat treatment process were further optimized based on the previous work. An enhanced transport J_c of 2.5 $\times$ 10^4 A/cm^2 at 4.2 K and in self-field was obtained in Fe-sheathed Sr-122 tapes with Sn metal additive and a heat treatment at 1100 °C for 5–15 min [13]. As summarized in Fig. 4.1, we can compare the transport J_c of Ba-122 tapes with Ag and Pb co-addition and Sr-122 tapes with Sn addition. Both of them were annealed at 1100 °C for up to 15 min. The best J_c values were achieved for Ba-122 tapes annealed for 5 min and for Sr-122 tapes annealed for 2.5 min, respectively. For Sr-122 tapes with Sn addition, though its best J_c value is higher than that for Ba-122 tapes with Ag and Pb co-addition, its J_c performance has a dramatic drop when the annealing time is longer than 5 min. Therefore, for 122-type iron-based superconductors with different metal additives, it can be concluded that the optimized parameters for heat treatment

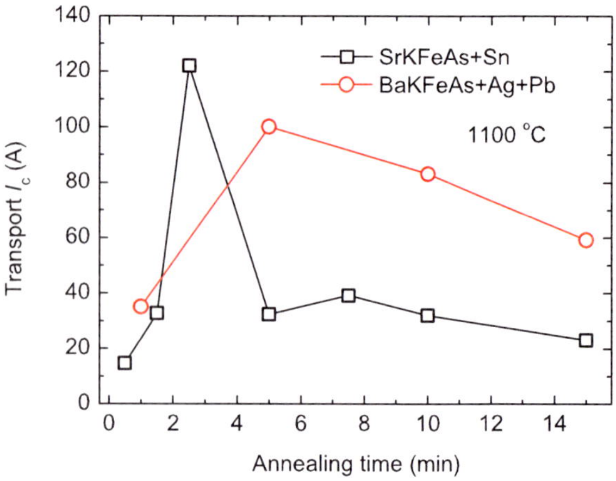

Fig. 4.1 Transport I_c at 4.2 K in self-field for Ba-122 tapes with Pb and Ag co-addition and Sr-122 tapes with Sn addition at 1100 °C with different annealing times

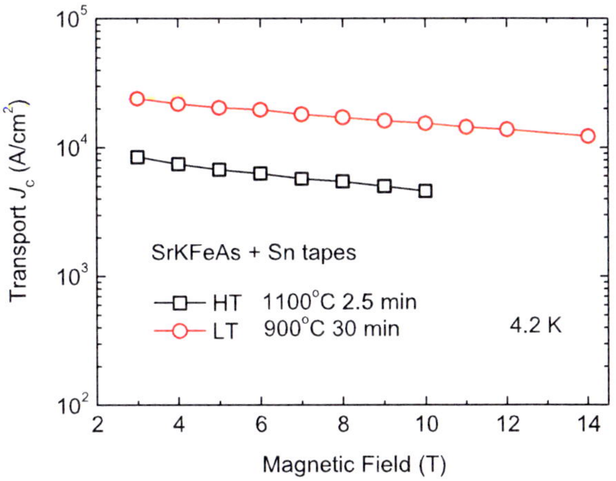

Fig. 4.2 Field dependence of the transport J_c of Sr-122 tapes with Sn addition annealed by using a short-time high-temperature (HT) and a low-temperature (LT) heat treatment. The applied field is parallel to the tape plane

would be also different. Through lowering the annealing temperature down to 800–950 °C, and moderately extending the annealing time to about 30 min, a much higher J_c performance (1.7×10^4 A/cm^2 at 4.2 K, 10 T) was achieved [14]. The field dependence of transport J_c for Sn added Sr-122 tapes annealed at 1100 °C for 2.5 min and at 900 °C for 30 min are compared in Fig. 4.2. Except for a significant enhancement of the J_c values, it can be noticed that the field dependence of J_c is weaker for the samples heat treated at 900 °C for 30 min, indicating a better crystallization of superconducting phase.

In order to further improving the J_c property, mechanical deformation process should be investigated and optimized. In the development of Bi-2223 tapes, the mass density and grain orientation were strengthened with the decreasing of tape thickness for flat rolling process, so nowadays the high-performance commercial Bi-2223/Ag tapes are usually about 0.2 mm or even smaller in thickness [15, 16]. The thickness of current Fe-sheathed Sr-122 and Ba-122 tapes are basically larger than 0.7 mm, so here Fe-sheathed Sr-122 tapes with various thicknesses were fabricated by flat rolling. First, the starting materials for synthesizing $Sr_{0.6}K_{0.4}Fe_2As_2$ precursor including small Sr pieces, K bulks and fine Fe and As powders were mixed and ground by ball milling in Ar atmosphere for about 12 h, then sealed and sintered in an Nb tube for 35 h at 900 °C. In order to compensate the loss of K during the sintering, an excess of 20% K was added. The precursor taken out of the Nb tube was added with 10 wt.% Sn to improve the grain connectivity, and then the mixture was ground into fine powder in an agate mortar in Ar atmosphere. The final powder was packed into Fe tubes (OD: 8 mm and ID: 5 mm), which were subsequently drawn into wires of 1.90 mm in diameter with a reducing rate of about 10%, and then flat rolled into tapes with different thicknesses. Finally, the tape samples were heat treated at 850–900 °C for 30 min under an Ar gas atmosphere. In order to investigate the effect of thickness on the Sr-122 tapes, they were flat rolled as thin as possible. However, due to the high hardness of iron and cold work hardening, starting from drawn wires of 1.9 mm in diameter, it is hard to roll the tapes thinner than 0.5 mm. The transport critical

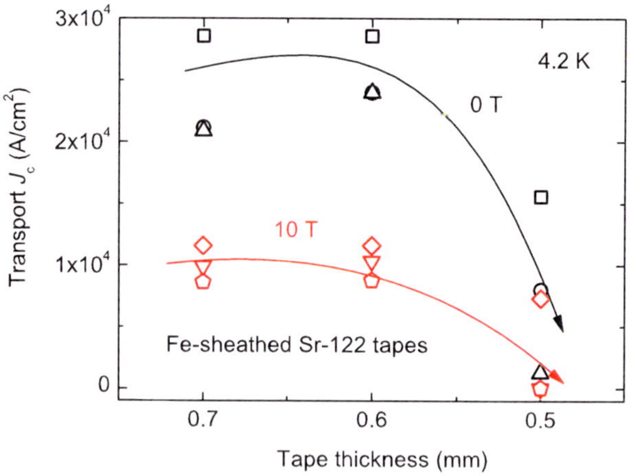

Fig. 4.3 Transport J_c of Fe-sheathed Sr-122 tapes flat-rolled into various thicknesses with Sn addition at 0 T and 10 T, respectively. The applied field is parallel to the tape plane

currents for the tapes with thicknesses of 0.7, 0.6 and 0.5 mm were then measured at 4.2 K in self-field and at 10 T, respectively. The corresponding J_c values are presented in Fig. 4.3. It can be seen that there is no big difference for the J_c values of 0.7 and 0.6 mm thick tapes, but there is a significant drop for the J_c values of 0.5 mm thick tapes. This may be caused by the large cracks in superconducting core induced by the deformation of hardened iron sheath. Therefore, a more extensive exploration for the cold defamation process and sheath materials for the Sr-122 tapes is needed for the next step.

4.2 Fe-Sheathed Tapes Processed with Uniaxial Pressing

Except for flat rolling, uniaxial pressing is another mechanical deformation method to achieve grain texture for Bi-2223 tapes. Compared with flat rolling, uniaxial pressing induces less micro cracks in grains during the mechanical deformation [17] and can achieve a higher mass density in superconducting cores [18]. Due to improved grain connectivity and high degree of texturing, Bi-2223 tapes processed with uniaxial pressing can obtain high transport J_c [19]. In addition, for flat rolling, the strength of deformation force is generally determined by the tape thickness and some parameters of rollers [20], so it is hard to quantitatively control the force during the deformation. In contrast, by applying different pressures for the pressing process, it is easier to investigate the relationship between the deformation force and the superconducting properties of tapes. Therefore, it is interesting to employ the cold uniaxial pressing for the fabrication of iron-based superconducting tapes.

In this work, Sr-122 tapes were fabricated with powder-in-tube (PIT) method and deformed directly from as-drawn wires by a one-step cold uniaxial pressing, as presented in Fig. 4.4. The process for synthesizing $Sr_{0.6}K_{0.4}Fe_2As_2$ precursor is the same as that introduced in Sect. 4.1. The obtained precursor was added with 10

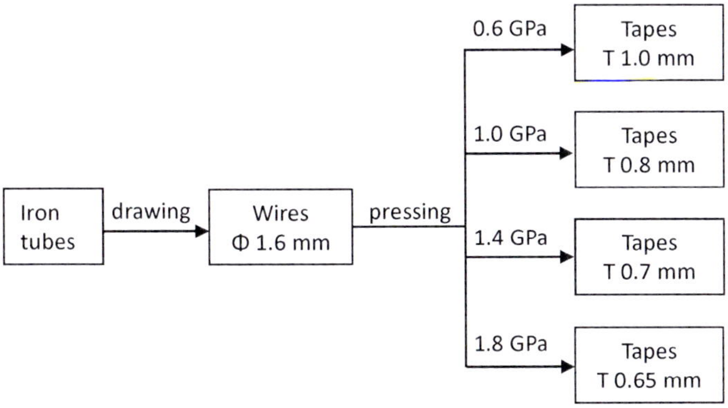

Fig. 4.4 Flow chart for the fabrication process of the Fe-sheathed Sr-122 tapes with various thicknesses cold pressed with various pressures

wt.% Sn to improve the grain connectivity, and then the mixture was ground into fine powder in an agate mortar in Ar atmosphere. The final powder was packed into Fe tubes (OD: 8 mm and ID: 5 mm), which were subsequently drawn into wires of 1.60 mm in diameter with a reducing rate of about 10%. The as-drawn wires were cut into short samples, and then pressed into tapes with maximal pressures of 0.6, 1.0, 1.4 and 1.8 GPa, respectively. Accordingly, the thickness of the pressed tapes is about 1.0, 0.8, 0.7 and 0.65 mm respectively, as shown in Fig. 4.5. The pressure on the tapes was held for about 3 min after the deformation. Finally, the as-pressed tapes were sealed into quartz tubes, heat treated at 850–950 °C for about 0.5 h. For comparison, the unpressed wire samples were also processed with the same heat treatment.

As shown in Fig. 4.6, when increasing the pressure of uniaxial pressing from 0.6 to 1.4 GPa, the transport J_c at 0 T and 5 T increases with the increasing pressure. However, when the pressure came up from 1.4 to 1.8 GPa, the transport J_c started to

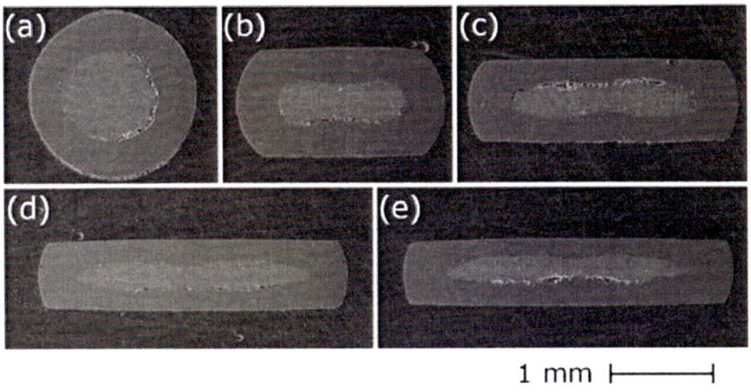

Fig. 4.5 SEM image of the transverse cross section of the **a** unpressed wire and tapes cold pressed at **b** 0.6 GPa, **c** 1.0 GPa, **d** 1.4 GPa and **e** 1.8 GPa. Reprinted from Yao et al. [21]

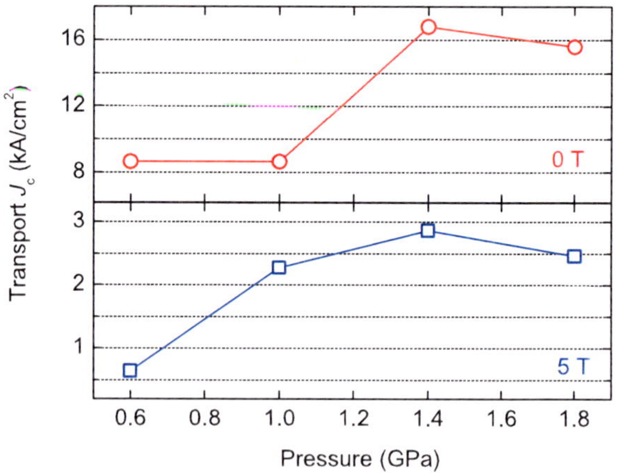

Fig. 4.6 Transport J_c (0 T) and J_c (5 T) at 4.2 K of Sr-122 tapes cold pressed at 0.6, 1.0, 1.4 and 1.8 GPa. The applied field is parallel to the tape plane

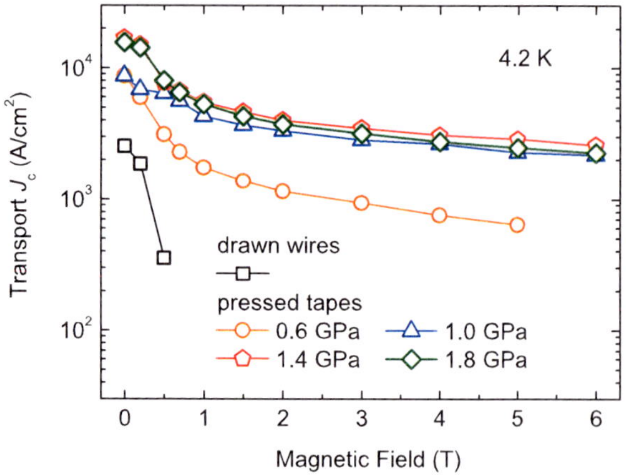

Fig. 4.7 Field dependence of the transport J_c (at 4.2 K) of the drawn Sr-122 wires and Sr-122 tapes pressed at 0.6, 1.0, 1.4 and 1.8 GPa. The applied field is parallel to the tape plane. Reprinted from Yao et al. [21]

decrease. The field dependent transport J_c of all the samples is presented in Fig. 4.7. Compared with that of unpressed tapes, the transport J_c of the samples pressed at 0.6 GPa is much higher in the whole field region, which is obviously the result of densified superconducting cores. When the pressure was incresed to 1.0 GPa, it can be seen that the transport J_c was enhcaned more significantly in high field region than in low field region, indicating that this J_c improvement can be mainly ascribed to the strengthened texture, which alleviates the weak-link effect of large-angle GBs. When the pressure was higher than 1.0 GPa, the change of transport J_c became small, implying small change in the mass density and grain alignment of superconducting cores. The highest transport J_c, which is obtained at 1.4 GPa, is up to 16.8 kA/cm^2 at 4.2 K in self field, and also exhibits weak field dependence in fields up to 6 T. For the tapes prcessed with a higher pressure of 1.8 GPa, there was a slight decrease for the J_c value, so further charcterizations are needed to find out the reason.

In order to analyze the grain texture, the superconducting cores of the tapes were examined by x-ray diffraction (XRD) after removing the Fe sheaths. For comparison,

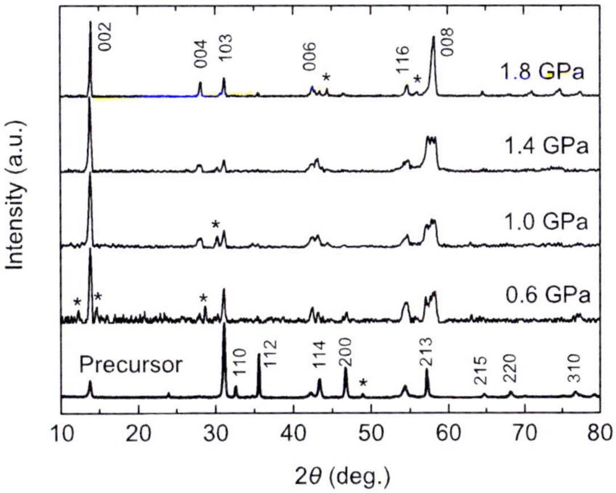

Fig. 4.8 XRD patterns for the precursor and the superconducting cores of Sr-122 tapes pressed at 0.6, 1.0, 1.4 and 1.8 GPa. The main peaks of Sr-122 phase are indexed, and the impurity phases are marked with the symbol*. Reproduced from Yao et al. [21]

the Sr-122 precursor was ground into fine powder and also submitted for XRD examination. The XRD patterns for the precursor and the superconducting cores of tapes processed with different pressure are shown in Fig. 4.8. Except some minor impurity peaks, all the diffraction patterns show strong peaks of Sr-122 phase, which are indexed in the figure. For a clearer comparison of the *c*-axis texture, the XRD data of the pressed tapes are normalized by the intensity of (002) peak. Compared with that of the precursor powder, it is obvious that the relative intensity of (002) peak with respect to that of (103) peak is strongly increased for all the pressed samples, indicating strong *c*-axis orientation induced by pressing. According to the Lotgering method [22], the degree of *c*-axis texture can be evaluated with an orientation factor $F = (\rho - \rho_0)/(1 - \rho_0)$, where $\rho = \Sigma I(00\ l)/\Sigma I(hkl)$, $\rho_0 = \Sigma I_0(00\ l)/\Sigma I_0\ (hkl)$, I and I_0 are the intensities of each reflection peak (hkl) for the textured and randomly oriented samples, respectively. The *c*-axis orientation factor F of the tapes pressed at 0.6, 1.0, 1.4 and 1.8 GPa is calculated to be 0.63, 0.72, 0.76 and 0.79 respectively, as presented in Fig. 4.9. It can be concluded that the degree of the grain texture

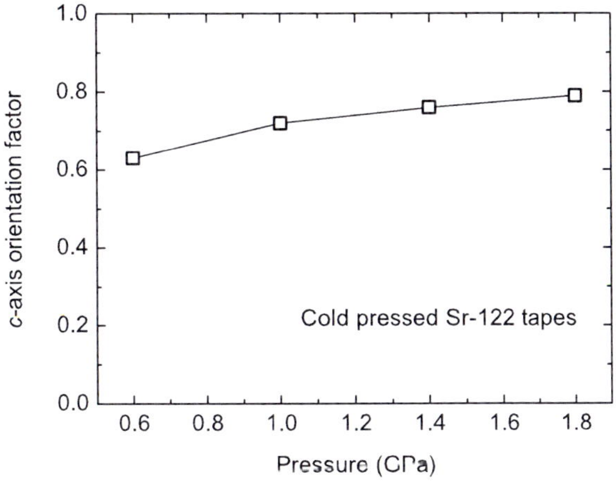

Fig. 4.9 Orientation factor F of *c*-axis texture for the Sr-122 tapes cold pressed at 0.6, 1.0, 1.4 and 1.8 GPa, respectively

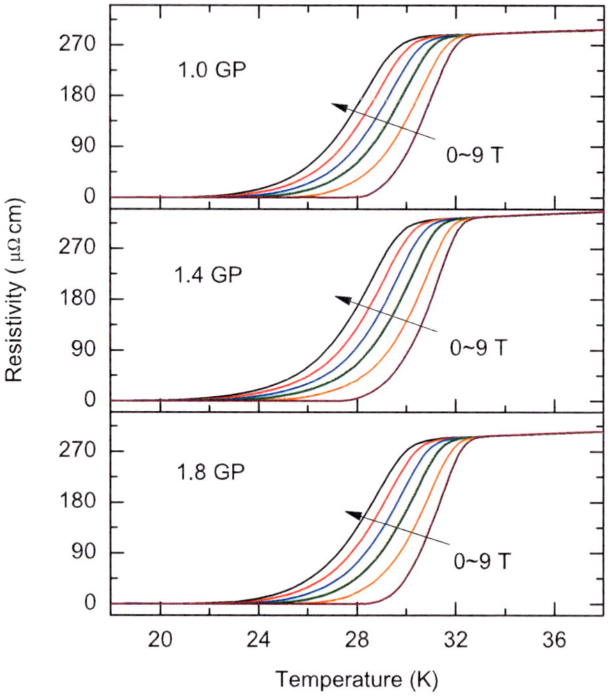

Fig. 4.10 Temperature dependence of resistivity at various magnetic fields for the Sr-122 tapes cold pressed at 1.0, 1.4 and 1.8 GPa, respectively

is enhanced with the increase of applied pressure. In addition, when increasing the pressure from 0.6 to 1.8 GPa, the cross-sectional areas of the Sr-122 core in each pressed sample are almost the same, suggesting that there is no further increase for the mass density of Sr-122 cores. Therefore, it can be confirmed that the main reason for the improvement of transport J_c when the pressure is higher than 0.6 GPa is the enhanced grain alignment.

The superconducting properties of the tapes cold pressed with different pressures are also systemically investigated. As presented in Fig. 4.10, the temperature dependence of the resistivity at 0–9 T was measured for the samples pressed at 1.0, 1.4 and 1.8 GPa, respectively. The *R-T* measurements of the Sr-122 cores are conducted after the Fe sheath of tapes was peeled off. The applied field is perpendicular to the surface of the Sr-122 cores. It can be seen that all the samples have sharp superconducting transitions, indicating good quality of Sr-122 phase after the heat treatment. The change of pressure has no obvious influence on the resistivity in normal state and its transition to superconducting state. The temperature dependence of the upper critical field H_{c2} and irreversibility field $H_{\rm irr}$, which are estimated with the criteria of 90% and 10% of resistivity at normal state respectively by using the Werthamer-Helfand-Hohenberg (WHH) formula (see Chap. 2 for detail), are shown in Fig. 4.11. There is also no big difference for the samples that were cold pressed at 1.0, 1.4 and 1.8 GPa. Therefore, it can conclude that when the pressure is above 1.0 GPa, the different pressures have no substantial influence on the superconducting properties

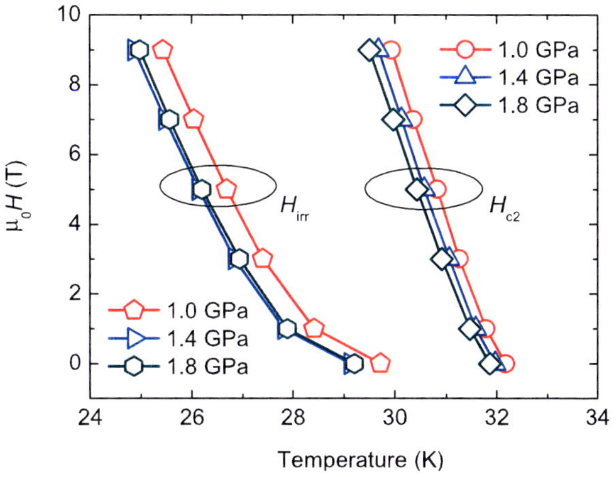

Fig. 4.11 Upper critical field $\mu_0 H_{c2}(T)$ and irreversibility field $\mu_0 H_{irr}(T)$ for for the Sr-122 tapes cold pressed at 1.0, 1.4 and 1.8 GPa, respectively

of the tapes, and the reason for the improvement of transport J_c is the enhanced grain texture.

In order to investigate the relationship between the microstructure and the transport properties of the pressed tapes, SEM observation was performed on the cross sections of Sr-122 cores for the samples pressed at 0.6, 1.0, 1.4 and 1.8 GPa, as shown in Fig. 4.12. In Fig. 4.12a, there are some planar structures in the Sr-122 core of the sample pressed at 0.6 GPa. However, as marked by arrows, due to the relatively low degree of texture, there are some randomly oriented grains. In addtion, some cracks were obseved for the Sr-122 phase. When the pressure is increased to 1.0 GPa, it is obvious that more cracks can be found, but they can be partially healed by the final heat treatment process. In Fig. 4.12c, in accordance with the improved grain texture which is comfirmed by the grain alignment obseved on the longitudinal cross senctions shown in Fig. 4.12e, f, the sample pressed at 1.4 GPa shows a well coupled planar structure with a few micro cracks. In contrast, in Fig. 4.12d, though smooth planar structure can also be observed, some unhealed cracks (marked by arrows) and crushed grains appear in the sample pressed at 1.8 GPa. During the mechanical deformation of cold pressing, the planar Sr-122 grians tend to be parallel to the the pressing plane under the deformation force, hence the c-axis texture is improved. On the other hand, some of the planar grains are crushed under high pressure. After the cold defornation, the final heat treatment can heal some micro cracks, and reconnect some crushed grains, but some serious cracks can not be healed by the heat treatment. Such residual cracks may counteract the positive effects on the transport properties brought by the improved texture. In this case, the transport J_c will decrease with the increase of pressure.

In conlusion, by optimizing the applied pressure, it is improtant to find a balance btween the texture and defects induced by the deformation force. In the future, if the heat treatment process can be properly adjusted to match the stronger pressing force, a higher transport J_c can be expected. On the other hand, due to the tapes in this work are directly pressed using the as-drawn wires, the freedom parameter for pressing

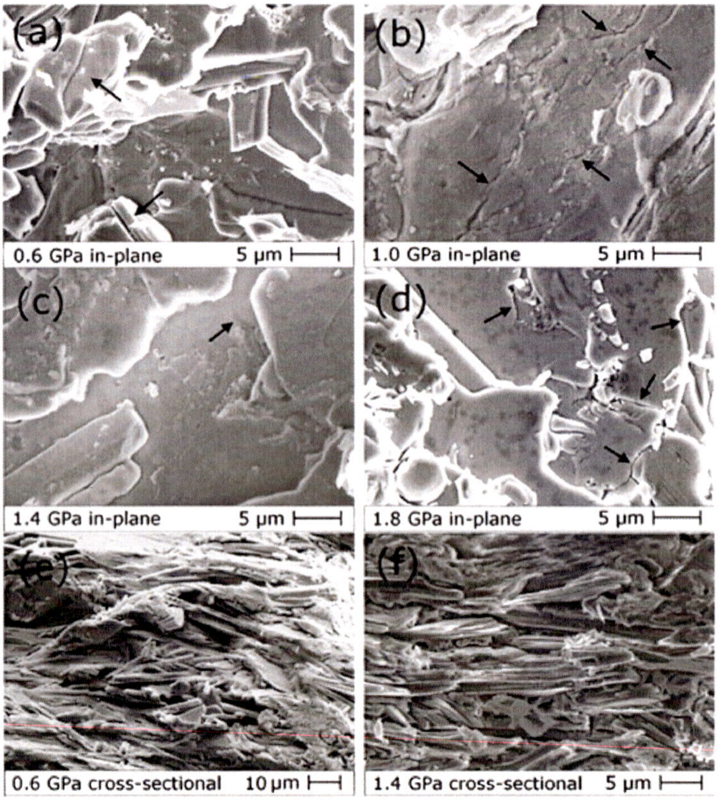

Fig. 4.12 SEM images of the superconducting cores, including the in-plane microstructure of the Sr-122 tapes cold pressed at **a** 0.6 GPa, **b** 1.0 GPa, **c** 1.4 GPa, **d** 1.8 GPa, and the cross-sectional microstructure of the tapes cold pressed at **e** 0.6 GPa and **f** 1.4 GPa. The isolated grains in (**a**) and micro cracks in (**b**), (**c**) and (**d**) are marked by arrows. Reprinted from Yao et al. [21]

$\Delta_f = h/L$ (where h is the thickness of the tape during the deformation and L is the contact length between the anvil and the tape in the width-direction) is relatively large, so the effect of densification is somewhat limited [20]. If the wires can be previously flat-rolled into tapes with a proper thickness, the freedom parameter for the pressing will become small. As a result, a higher mass density and better grain tecture can be achieved.

4.3 Ag-Sheathed Tapes Processed with Flat Rolling

Besides the iron sheath used in the above section, silver is also another sheath material suitable for the 122-type iron-based superconducting wires and tapes. Compared with iron, silver is much weaker in mechanical strength, but on the other hand it is more ductile for cold-work deformation and more chemically stable during the heat treatment process for the fabrication of iron-pnictide wires and tapes. However, due to the lower melting point of silver (~961 °C), the high-temperature annealing at 1100 °C that used in Chap. 3 cannot be used for silver sheathed tapes. Nevertheless, as shown in Sect. 4.2, an improved transport J_c has been obtained by low-temperature annealing at around 900 °C, which can also be applied to silver sheathed tapes. With the use of ductile silver sheath, it is expected that the micro cracks can be suppressed and the thickness of tapes can be further optimized for achieving a better J_c performance.

In this work, the preparation of Sr-122 precursor is similar to that in Sect. 4.2. The starting materials for synthesizing $Sr_{0.6}K_{0.4}Fe_2As_2$ precursor are small Sr pieces, K bulks and fine Fe and As powders. In order to compensate the loss of K during the sintering, an excess of 20% K was added. The powders were mixed and ground by ball milling in Ar atmosphere for about 12 h, then sealed and sintered in an Nb tube for 35 h at 900 °C. The sintered precursor was added with 5 wt. % Sn and ground into fine powder in an agate mortar in Ar atmosphere. The final powder was packed into Fe tubes (OD: 8 mm and ID: 5 mm), which were subsequently drawn into wires of 1.9 mm in diameter with a reducing rate of about 10%. Then the wires were rolled into flat tapes with thicknesses of 0.6, 0.5, 0.4, and 0.3 mm, respectively, as shown in Fig. 4.13. The as-drawn wires and as-rolled tapes were sealed into quartz tubes and finally heat treated at 900 °C for about 0.5 h. The transverse cross sections observed with optical microscope of silver sheathed Sr-122 wire with 1.87 mm in diameter and tapes with 0.6 and 0.4 mm in thickness are presented in Fig. 4.14.

The magnetic field dependences of the transport J_c for the drawn Sr-122 round wire and rolled Sr-122 tapes with various thicknesses are shown in Fig. 4.15. The transport J_c of the drawn wire reaches 10^4 A/cm^2 at 4.2 K in self-field. However, it drops rapidly by one order of magnitude when the applied field was increased to 4 T. This result is similar to the Ba-122/Ag grooved rolled wires reported by Togano et al. [24], indicating a weak grain connectivity in the Sr-122 superconducting core.

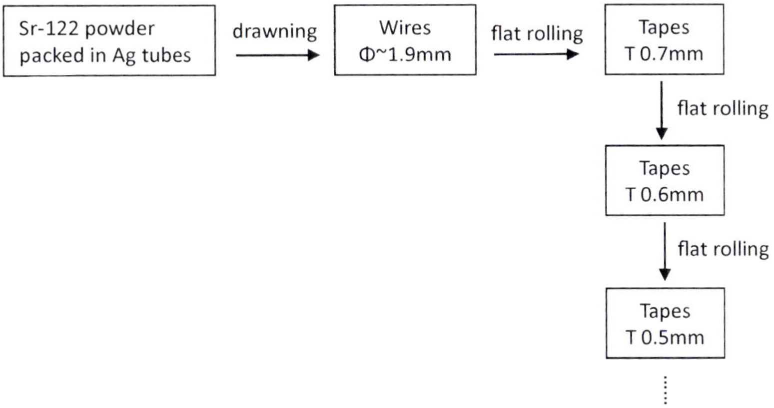

Fig. 4.13 Flow chart for the fabrication process of the flat rolled Ag-sheathed Sr-122 tapes with various thicknesses

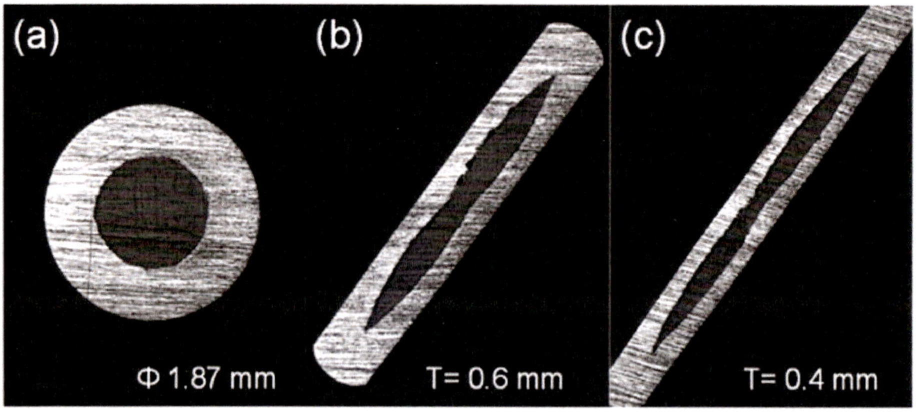

Fig. 4.14 Optical images of the transverse cross section for the Ag-sheathed Sr-122 wires and tapes. **a** Drawn wire of 1.87 mm in diameter; **b** flat rolled tapes of 0.6 mm in thickness; **c** flat rolled tapes of 0.4 mm in thickness

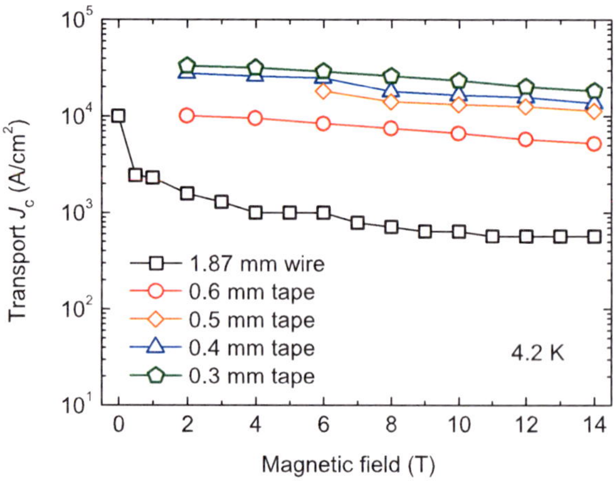

Fig. 4.15 Field dependence of the transport J_c (at 4.2 K) of the drawn Sr-122 wires and Sr-122 tapes with various thicknesses. The applied field is parallel to the tape plane. Reprinted from Yao et al. [23]

In contrast to the wires, because of the rolling induced densification of the superconducting cores for the tapes, it can be seen that there is significant improvement in transport J_c, especially for its performance in the high-field region. For the tapes with different thicknesses, the transport J_c in the whole field region can be further increased by rolling with smaller thicknesses. The highest J_c at 4.2 K and 10 T reaches 2.3×10^4 A/cm^2 in tapes of 0.3 mm in thickness. This J_c value is about four times higher than that for the 0.6 mm thick tape. The relationship between the J_c performance and the thickness of tapes is summarized in Fig. 4.16, which shows a roughly linear correlation between them. Therefore, the transport J_c of Sr-122 tapes can be effectively increased by high-reduction-ratio rolling process.

The XRD patterns of the superconducting cores examined after removing the Ag sheath for rolled Sr-122 tapes with 0.3 and 0.6 mm in thickness are shown in Fig. 4.17. The XRD data of the tapes are normalized by the intensity of the (002)

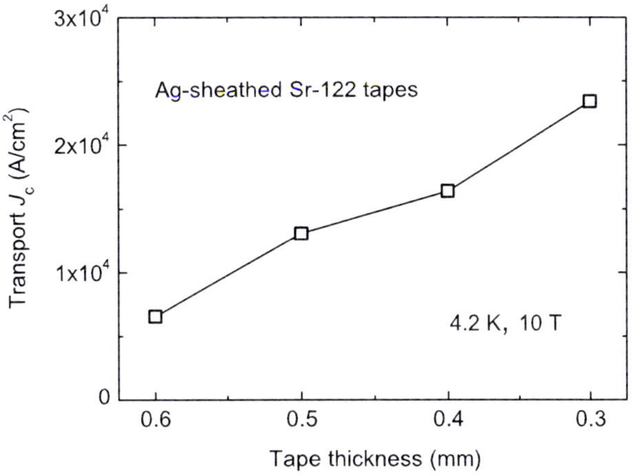

Fig. 4.16 Evolution of the transport J_c (at 4.2 K and 10 T) of the flat rolled Sr-122 tapes with various thicknesses. The applied field is parallel to the tape plane

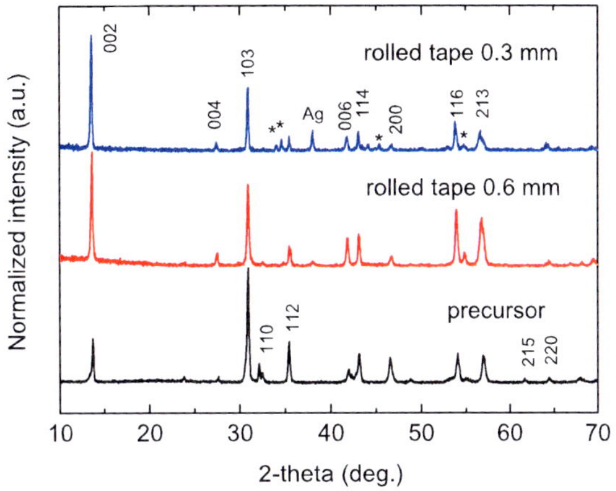

Fig. 4.17 XRD patterns for the precursor and the superconducting cores of the Sr-122 tapes of 0.6 and 0.3 mm in thickness. The main peaks of Sr-122 phase are indexed, and the impurity phases are marked with the symbol*

peak. It can be seen that except for the minor Ag peaks due to the sheath material, all the diffraction patterns show strong peaks of the Sr-122 phase, which are indexed in the figure. In addition, the XRD pattern for the precursor powder is also shown for comparison. Compared to the XRD pattern of the precursor powder, it is obvious that the relative intensities of the (002) peaks with respect to that of the (103) peaks are strongly increased for all the rolled samples, indicating a strong *c*-axis grain orientation. Furthermore, the relative intensities of the (002) peaks for the 0.3 mm thick tapes are higher than that for the 0.6 mm thick tapes, indicating an improved *c*-axis texture, which is considered as the main reason for the enhancement of J_c in the 0.3 mm thick tapes. For the quantitative characterization for the grain texture, the degree of grain texture can be evaluated with an orientation factor (F) by the Lotgering method [22] $F = (\rho - \rho_0)/(1 - \rho_0)$, where $\rho = \Sigma I(00l)/\Sigma I(\text{hkl})$, $\rho_0 = \Sigma I_0(00l)/\Sigma I_0(\text{hkl})$, I and I_0 are the intensities of each reflection peak (hkl) for the textured and randomly oriented samples, respectively. The F values of the tapes with 0.6 and 0.3 mm in thickness is 0.30 and 0.40, respectively. It can be concluded

Fig. 4.18 Temperature dependence of normalized magnetization of the Sr-122 tapes of 0.6 and 0.3 mm in thickness measured in zero-field (ZFC) and field cooling (FC) procedures

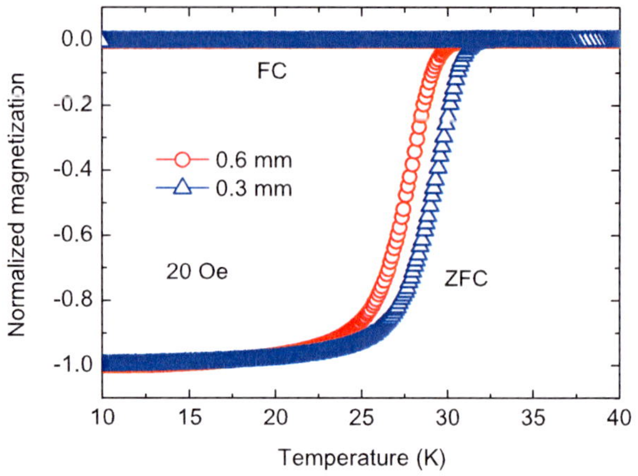

that the degree of grain texture can be increased by rolling the tape with a smaller thickness.

Figure 4.18 shows the temperature dependence of susceptibility of the superconducting cores in Sr-122 tapes with 0.6 and 0.3 mm in thickness with zero field cooling (ZFC) and field cooling (FC) process under a 20 Oe field. It can be seen that the onset transition temperature of the 0.3 mm thick tapes is higher than that of the 0.6 mm thick tapes. This is in accordance with the results of resistivity measurement shown in Fig. 4.19. The increased transition temperature of 0.3 mm thick tapes can be ascribed to the enhanced mass density of the Sr-122 superconducting cores by further rolling,

Fig. 4.19 Temperature dependence of resistivity at various magnetic fields for the Sr-122 tapes of 0.6 and 0.3 mm in thickness, respectively

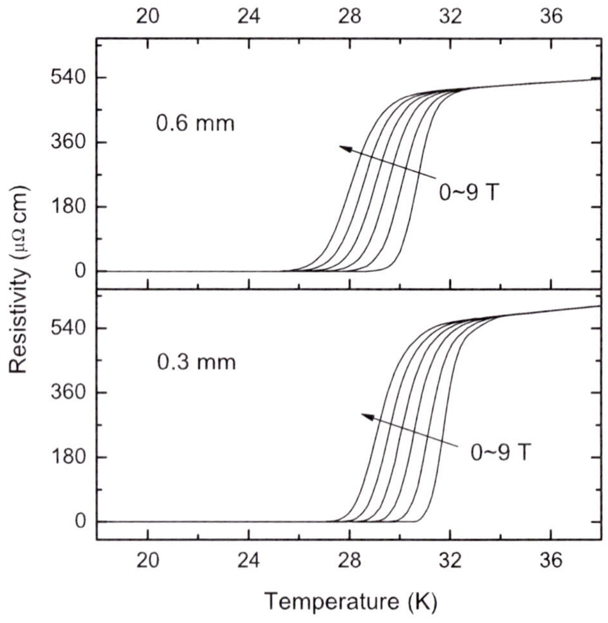

which promoted the phase formation of Sr-122 material. In Fig. 4.19, it can be noted that except for the higher transition temperatures, the expansion of *R-T* curves with the increasing fields towards low-temperature region is also smaller for the 0.3 mm thick tapes than that of the 0.6 mm thick tapes, indicating an enhanced flux pinning behavior. Such improved superconducting properties in the 0.3 mm thick tapes are also beneficial to the J_c performance.

4.4 Ag-Sheathed Tapes Processed with Uniaxial Pressing

Compared with the Fe-sheathed Sr-122 tapes in Sect. 4.1, the transport J_c of the Ag-sheathed Sr-122 tapes in Sect. 4.3 is higher because of the higher reduction ratio for the cold rolling process. However, due to the softness of sheath material, the grain orientation factor $F = 0.40$ for the 0.3 mm thick Ag-sheathed tapes is even lower than 0.42 for the Fe-sheathed tapes with 0.6 mm in thickness. In Sect. 4.2, it is shown that the grain orientation can be greatly enhanced by introducing uniaxial pressing process. Therefore, much higher-degree grain texture can be also expected for the Ag-sheathed tapes by cold pressing. In order to alleviate the cold-work hardening of the Ag sheath and reduce the deformation induecd cracks inside the superconducting cores as observed in the pressed Fe-sheathed tapes, an intermediate heat treatment process was added between the cold rolling and cold pressing processes. On the other hand, hot pressing process, which conbines the pressing with the final heat treatment in a single procedure, was applied to as rolled Sr-122 tapes, as illustrated in Fig. 4.20. With these new fabrcating routes, well-textuted and highly-dense Sr-122 tapes with higher J_c performance can be expected.

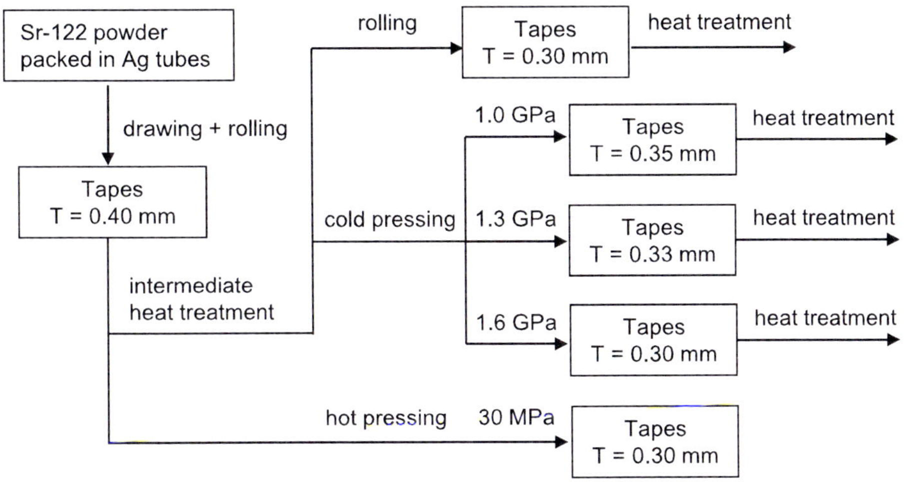

Fig. 4.20 Flow chart for the fabrication process of the Ae-sheathed Sr-122 tapes with flat rolling, cold pressing combined with an intermediate heat treatment, and one-step hot pressing

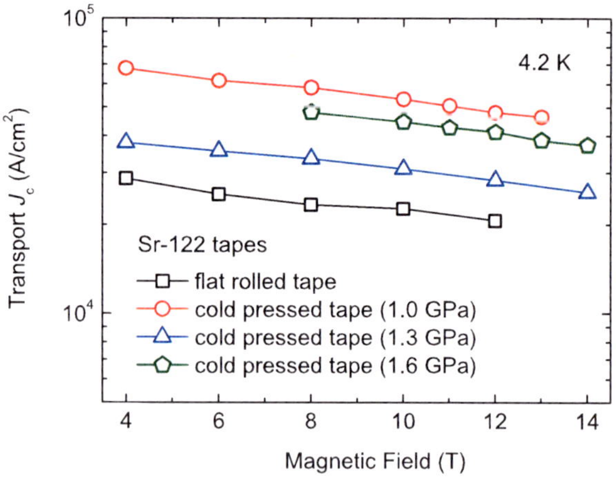

Fig. 4.21 Field dependence of the transport J_c (at 4.2 K) of the Ag-sheathed Sr-122 tapes fabricated with different thermomechanical processes. The applied field is parallel to the tape plane

The Sr-122 precursor powder used in this section is the same as the one used in Sect. 4.3. The Sr-122 powder was first packed into Ag tubes with an OD of 8 mm and ID of 5 mm, which were subsequently drawn into wires of ~1.9 mm diameter and rolled into flat tapes with 0.40 mm in thickness. The as-rolled tapes were submitted to an intermediate heat treatment at 850 °C for about 30 min, and then uniaxially pressed between two steel dies under a pressure of about 1.0, 1.3 and 1.6 GPa, respectively. The thickness of the as-pressed tapes is 0.35, 0.33 and 0.30 mm, respectively. For the comparison between rolling and pressing processes, another 0.30 mm thick tape was fabricated by using rolling for the first and second cold-work deformation processes, as shown in Fig. 4.20. Finally, all the tape samples were submitted to a final heat treatment at 900 °C for about 30 min. For the hot pressed samples, the as rolled tapes were sandwiched between two pieces of metal sheets and pressed at ~30 MPa at 850 °C for 30 min.

The field dependence of the transport J_c all the Sr-122 tape samples are presented in Fig. 4.21. The transport J_c of the roll-HT-roll-HT tape is 2.3×10^4 A/cm^2 at 4.2 K, 10 T, which is on the same level of tapes with the same thickness made by the conventional rolling process. For the tapes made by the roll-HT-press-HT process, the J_c performace were significantly improved. The highest J_c 4.8×10^4 A/cm^2 (4.2 K, 10 T) was achieved in the tapes pressed at 1.0 GPa, and this J_c value is more than two times higher than that for the tapes made without cold pressing. However, with the increase of pressure, the dur to the pressure induced cracks inside the superconducting cores, the transport J_c was suppressed, similar to the Fe-sheathed tapes in Sect. 4.2. Though the J_c for the tapes pressed with the highest pressure 1.6 GPa drops to 3.1×10^4 A/cm^2 (4.2 K, 10 T), it is still higher than that for the tapes made without cold pressing. In general, the uniaxial pressing is more effective in improving the J_c performance than flat rolling.

The XRD patterns of the superconducting cores for the intermediate-annealed Sr-122 samples with and without cold pressing are presented in Fig. 4.22. The XRD data of all the tapes are normalized by the intensity of the (103) peak. It can be seen

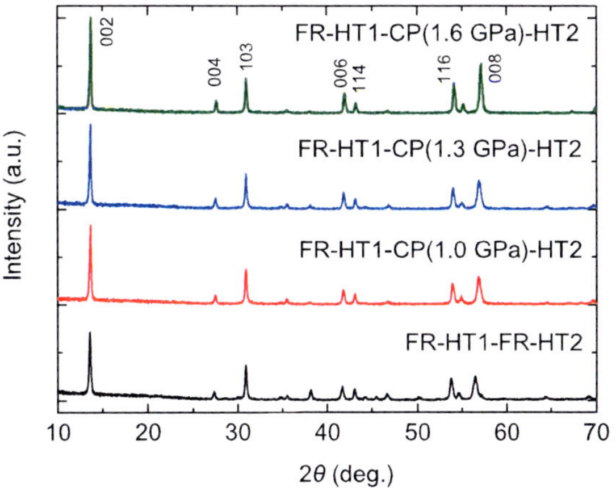

Fig. 4.22 XRD patterns for the Ag-sheathed Sr-122 tapes fabricated with flat rolling process and rolling-pressing process. The main peaks of Sr-122 phase are indexed. FR: flat rolling; CP: cold pressing (with various pressure of 1.0, 1.3 and 1.6 GPa); HT1: intermediate heat treatment; HT2: final heat treatment

that all the diffraction patterns show strong (00 *l*) peaks of the Sr-122 phase, which mean well *c*-axis textured grains. With the Lotgering method [22], the calculated orientation factor *F* for the roll-HT-roll-HT tape is 0.43, which is similar to the tapes with the same thickness made by the conventional rolling process. The *F* value for the cold pressed tapes with 1.0, 1.3 and 1.6 GPa pressure is 0.44, 0.48 and 0.50, respectively. These results show that the c-axis texture for the intermediate-annealed Sr-122 samples is stronger than that for the tapes made by the conventional rolling process in Sect. 4.3, and the texture can be further strengthened by cold pressing. With the increase of the applied pressure, the degree of texture gradually grows, but in contrast the J_c performance will degrade when the pressure is too high, which is similar to the result of Fe-shaethed tapes.

According to the results of XRD characterization, it can be confirm that the cold pressing proicess is very effective in improving the grain texture for the Sr-122 tapes. On the other hand, the mass density of the Sr-122 supercondcuting core can also be greatly enhanced by cold pressing, and this is considered to be the reason why the J_c of 0.3 mm thick tapes made by roll-HT-press-HT process is much higher than that of the tapes made by roll-HT-roll-HT process. Figure 4.23 presents the microstructure of longitudinal cross-section of the superconducting cores observed through SEM for the rolled and pressed tapes. Many planar Sr-122 grains can be seen in these two samples, but compared to the rolled tape, the grain alignment is more uniform in the pressed tape, which is in agreement with the XRD results. Moreover, with fewer pores, the Sr-122 phase of the pressed tape is much more dense than that for the rolled tape. With the characterization mentioned above, it is clear that cold pressing can improve the mass density, grain texture and formation of superconducting phase, so a higher transport J_c can be achieved in pressed tapes than in rolled tapes.

The magnetic field dependence of the J_c of flat rolled and hot pressed Sr-122 tapes are presented in Fig. 4.24. For the flat rolled tapes, J_c achieved a maximum value of 3.0×10^4 A/cm^2 at 10 T, 4.2 K. However, after the pressing, the J_c of Sr-122 sample is shown to be over the practical level of 10^5 A/cm^2 in 10 T at 4.2 K, and still remain 8.4×10^4A/cm^2 up to 14 T. As a reference, the J_c data for conventional NbTi and

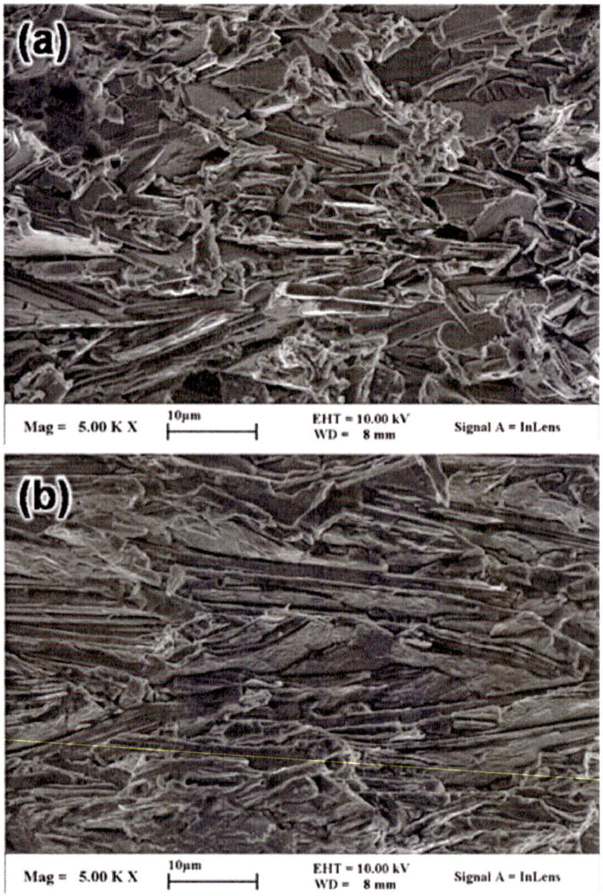

Fig. 4.23 SEM images for the longitudinal cross sections of the superconducting cores of the Ag-sheathed Sr-122 tapes fabricated with **a** flat rolling process and **b** rolling-pressing process (cold pressed at 1.0 GPa). Reprinted from Yao et al. [23]

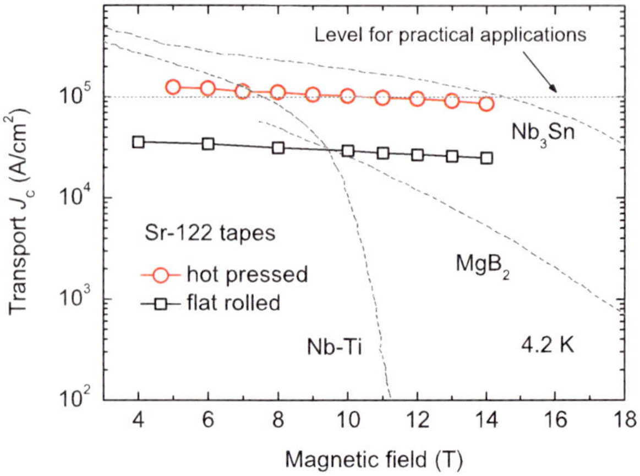

Fig. 4.24 Field dependence of transport J_c values at 4.2 K for the flat rolled and hot pressed Sr-122 tapes. The magnetic field was applied parallel to the tape plane

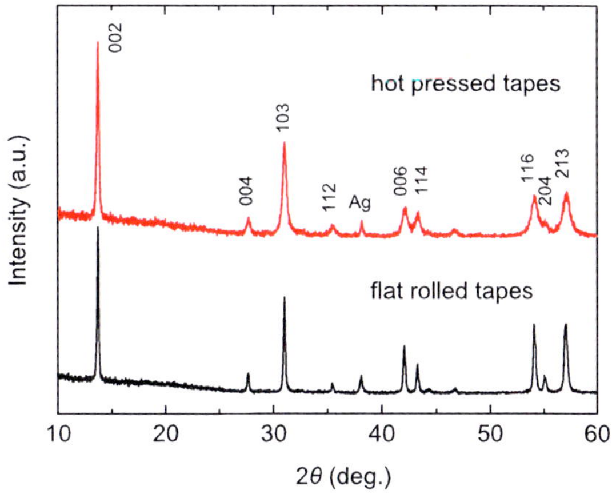

Fig. 4.25 XRD patterns for the Ag-sheathed Sr-122 tapes prepared by flat rolling and hot pressing. The main peaks of Sr-122 phase are indexed

Nb_3Sn superconducting wires and the PIT-processed MgB_2 wires were also included in the figure [25]. Notably, the hot pressed Sr-122 tapes have a superior J_c than that for MgB_2 and Nb-Ti in field region over 10 T, showing very weak field dependence of J_c, which is a desirable character for high-field applications.

Figure 4.25 shows XRD patterns of the tapes processed by rolling and pressing. Obviously, both of the samples exhibit a well-defined $ThCr_2Si_2$-type crystalline structure except the Ag peak which is ascribed to the Ag sheath. The XRD data indicate that a single-phase $Sr_{1-x}K_xFe_2As_2$ superconductor was obtained. The degree of the grain alignment can be estimated by the Lotgering method using the results of XRD examination [22]. Between the rolled and pressed samples, stronger relative intensity of (00 l) peaks was observed in the pressed tape. The c axis orientation factors F calculated by the Lotgering method for rolled and pressed samples are 0.45 and 0.52, respectively. Compared to the results of cold pressed tapes, a higher degree of texture was obtained in these samples due to a larger pressure applied during the hot pressing. This is in consistent with the previous textured Bi-2223 tapes [26], where hot pressing could enhance J_c by improving the grain alignment.

The typical SEM images of superconducting cores in flat rolled and hot pressed Sr-122/Ag tapes are shown in Fig. 4.26. Compared to the low-density structure of rolled tapes in Fig. 4.26a, a much denser Sr-122 core without voids and cracks can be seen in Fig. 4.26b for hot pressed tapes. This is similar to that happened in the Bi-2223 tape, where the pressure applied during heat treatment is very effective in the reduction of porosity as well as impurity phases [27]. In the out-of-plane longitudinal section, clear presence of uniformly textured grains along the tape axis is can be observed as shown in Fig. 4.26c. It is believed that the rolling and pressing process promotes grain alignment with the c-axis perpendicular to the tape surface because the planar Sr-122 crystals can be easily cleaved along the ab-plane. In the enlarged part in Fig. 4.26c, bent grains without cracks were found, which can be ascribed to the plastic deformation during hot press process, indicating that this technique can make the grains more flexible to couple with each other without producing a large

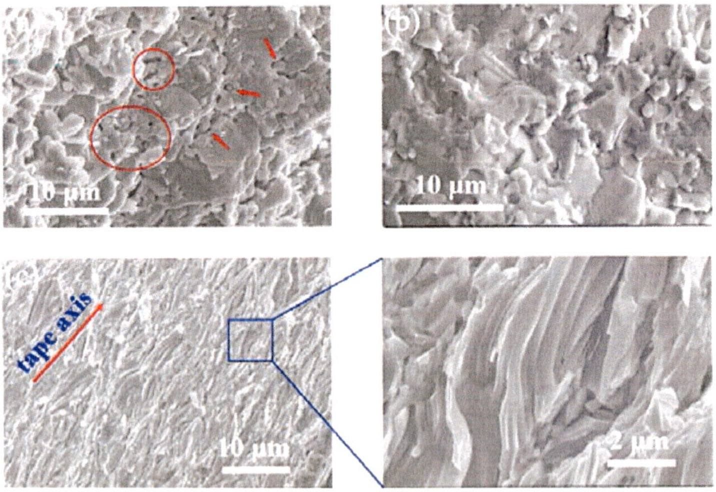

Fig. 4.26 Microstructure of the superconducting core of Sr-122 tapes investigated by SEM. **a** Flat rolled samples viewed from the tape plane direction. Some voids and cracks were indicated by arrows or circles. **b** Pressed samples viewed from the tape plane direction. **c** Hot pressed samples viewed from the longitudinal cross-sections. Reprinted from Zhang et al. [29]

number of crashed grains. It was reported that there were residual cracks along the tape axis in cold pressed Ba-122 superconducting tapes [28]. However, such crack is absent in our hot pressed Sr-122 samples. This suggests that hot deformation can help in effectively eliminating defects effectively while densifying the superconducting cores.

Figure 4.27 presents the magnetization versus temperature (M-T) curves for the superconducting cores of flat rolled and uniaxial pressed tapes under zero field cooling and field cooling. Their onset $T_c s$ are 34.4 K, 33.8 K and 34.2 K for tapes made by flat rolling, cold pressing and hot pressing, respectively. There is a slight

Fig. 4.27 Temperature dependence of normalized magnetization of the Sr-122 tapes prepared with flat rolling, cold pressing and hot pressing processes measured in zero-field (ZFC) and field cooling (FC) procedures

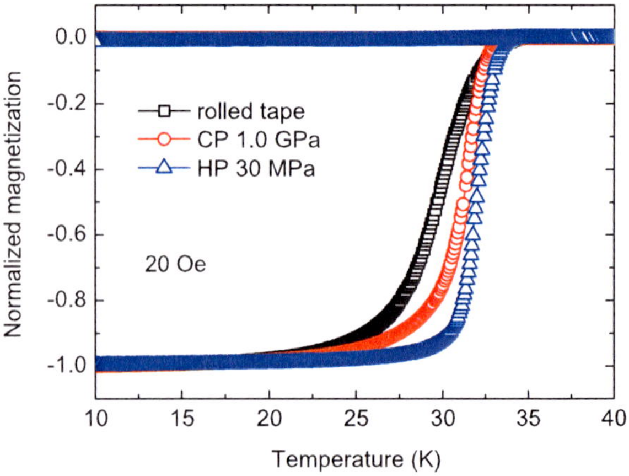

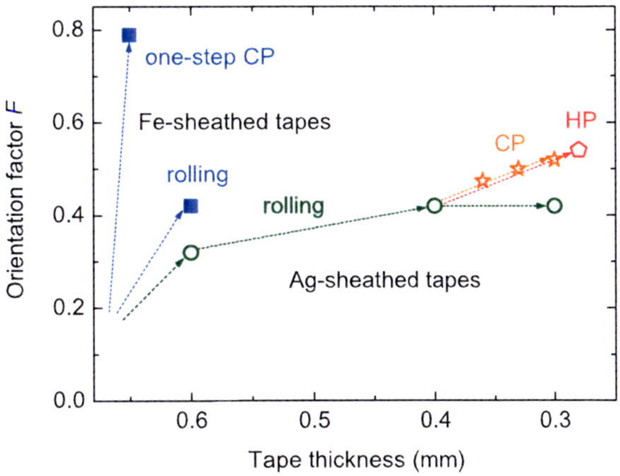

Fig. 4.28 Relationship between the *c*-axis orientation factor *F* and the thickness of tapes with different metal sheaths and fabricated by different processes

degradation of T_c after the cold and hot pressing. The T_c is the lowest for cold pressed samples, which may be ascribed to the loss of elements such as K and As during the intermediate heat treatment. However, the superconducting transition of the cold pressed tapes became much steeper than the rolled tapes, indicating that the pressing can promote the formation of the Sr-122 phase. The *M-T* curve for hot pressed samples shows a temperature-independent magnetic moment at low temperatures and a sharp superconducting transition, underlining the high quality of the samples. The hot pressed samples show a stronger diamagnetic signal than the cold pressed one, indicating that a high-quality superconducting phase was formed after the hot pressing.

For a more detailed discussion for the grain texture, the grain orientation factor *F* for the Fe-sheathed and Ag sheathed Sr-122 tapes made by flat rolling and uniaxial pressing is summarized in Fig. 4.28. It is clear that cold pressing is more powerful for improving the texture than rolling. For example, the *F* value for the Fe-sheathed tapes processed with one-step pressing at 1.8 GPa is as high as 0.79, nearly two times higher than that for the tapes made by flat rolling, though the thickness of rolled tapes is even smaller. For the Ag-sheathed tapes, when the tapes thickness was reduced smaller than 0.4 mm, the grain texture can hardly gain any further improvement by rolling, which can be ascribed to the weak mechanical strength of silver. Compare with the Ag sheath, the Fe sheath is more effective in achieving high-degree grain texture by using either flat rolling or uniaxial pressing process. However, the high stress during the deformation of Fe-sheathed tapes resulted in micro cracks which cannot be eliminated by heat treatment, and thus seriously worsen the grain connectivity. For further optimizing the grain connectivity and grain orientation, cold pressing combined with intermediate annealing and one-step hot pressing were utilized for Ag sheathed tapes, and further improvement on grain texture was achieved. Based on these result, to enhance the mechanical strength of Ag sheath, a reinforcing outer sheath made of stiff metal such as iron, stainless steel and copper can be added to form a composite sheath. The inner Ag sheath can buffer the severe deformation stress and strain from the outer sheath, thus reducing the cracks in the superconducting cores.

With the composite sheath, an excellent balance between the mechanical deformations induced densification and texture and the cracks caused by deformation stress, can be realized by synergetic microstructural tailoring, which is the key point to achieve superior J_c in the future.

4.5 Conclusion

In summary, based on an ex-situ powder-in-tube method, we fabricated Fe-sheathed Sr-122 superconducting tapes by flat rolling and cold uniaxial pressing, respectively. It is found that the transport J_c reaches a maximum value at the thickness of 0.6 mm by flat rolling. On the other hand, the as-drawn wires were directly pressed into tapes with various pressures of 0.6, 1.0, 1.4 and 1.8 GPa. It was found that the pressing can improve the mass density and induce c-axis texture in Sr-122 cores. Moreover, the c-axis texture of the pressed tapes can be further improved by gradually increasing the applied pressure, but the improvement of transport J_c stops at 1.8 GPa. With the investigation on the evolution of microstructure of Sr-122 cores processed with various pressures, it is found that in addition to enhancing the texture, the strong pressing force during the mechanical deformation can also induce crushed grains and cracks in the superconducting core, which will deteriorate the transport current. By using Ag sheath instead of Fe sheath, Sr-122 tapes were fabricated with conventional multi-step rolling process and also a combined process including rolling, intermediate annealing and uniaxial pressing. It is found that pressing is more effective in improving the grain texture when applied to tapes with smaller thickness, while when the tapes thickness was reduced smaller than 0.4 mm, the grain texture can hardly gain any further improvement by rolling. A high transport J_c of 4.8×10^4 A/cm^2 (4.2 K, 10 T) was obtained in pressed Sr-122/Ag tapes with 0.35 mm in thickness. Meanwhile, by using hot pressing technique, which combines the uniaxial press deformation with the heat treatment in one procedure, a practical-level J_c of 10^5 A/cm^2 at 4.2 K and 10 T was obtained for the first time. The comparative study of rolled and pressed tapes revealed that pressing can further improve the core density and grain texture of the superconducting phase in the as-rolled tapes, and thus achieving superior J_c performance.

References

1. Durrell JH, Eom CB, Gurevich A, Hellstrom E, Tarantini C, Yamamoto A, Larbalestier DC (2011) The behavior of grain boundaries in the Fe-based superconductors. Rep Prog Phys 74:124511
2. Katase T, Ishimaru Y, Tsukamoto A, Hiramatsu H, Kamiya T, Tanabe K, Hosono H (2011) Advantageous grain boundaries in iron-pnictide superconductors. Nat Commun 2:409
3. Lee S, Jiang J, Weiss JD, Folkman CM, Bark CW, Tarantini C, Xu A, Abraimov D, Polyanskii A, Nelson CT, Zhang Y, Baek SH, Jang HW, Yamamoto A, Kametani F, Pan XQ, Hellstrom

EE, Gurevich A, Eom CB, Larbalestier DC (2009) Weak-link behavior of gran boundaries in superconducting $Ba(Fe_{1-x}Co_x)_2As_2$ bicrystals. Appl Phys Lett 95:212505

4. Sakagamia A, Kawaguchia T, Tabuchib M, Ujiharac T, Takeda Y, Ikuta H (2013) Critical current density and grain boundary property of $BaFe_2(As, P)_2$ thin films. Physica C 494:181
5. Yamada Y, Watanabe T, Muroga T, Miyata S, Iwai H, Ibi A, Shiohara Y, Katoh T, Hirayama T (2005) Rapid production of buffered substrates and long length coated conductor development using IBAD, PLD methods and "Self-Epitaxial" ceria buffer. IEEE Trans Appl Supercond 15:2600
6. Aytug T, Paranthaman M, Zhai HY, Leonard KJ, Gapud AA, Thompson JR, Martin PM, Goyal A, Christen DK (2005) Iridium: an oxygen diffusion barrier and a conductive seed layer for RABiTS-based coated conductors. IEEE Trans Appl Supercond 15:977
7. Rupich MW, Schoop U, Verebelyi DT, Thieme C, Zhang W, Li X, Kodenkandath T, Nguyen N, Siegal E, Buczek D, Lynch J, Jowett M, Thompson E, Wang JS, Scudiere J, Malozemoff AP, Li Q, Annavarapu S, Cui S, Fritzemeier L, Aldrich B, Craven C, Niu F, Schwall R, Goyal A, Paranthaman M (2003) YBCO coated conductors by an MOD/RABiTSTM process. IEEE Trans Appl Supercond 13:2458
8. Ko RK, Park C, Kim HS, Chung JK, Ha HS, Shi DQ, Song KJ, Yoo SI, Moon SH, Kim YC (2005) Fabrication of meter-long coated conductor using RABiTS-PVD methods. IEEE Trans Appl Supercond 15:2707
9. Grasso G, Hensel B, Jeremie A, Flukiger R (1995) Distribution of the transport critical-current density in ag sheathed $(Bi, Pb)_2Sr_2Ca_2Cu_3O_x$ tapes produced by rolling. Physica C 241:45
10. Merchant N, Luo JS, Maroni VA, Riley GN Jr, Carter WL (1994) Reaction induced texture of $(Bi, Pb)_2Sr_2Ca_2Cu_3O_{10+\delta}$/Ag composite. Appl Phys Lett 65:1039
11. Hu QY, Liu HK, Dou SX (1996) Effect of mechanical deformation on the mass density of Ag-clad $(Bi, Pb)_2Sr_2Ca_2Cu_3O_{10}$ wire and tape. Appl Supercond 4:17
12. Wang L, Qi Y, Zhang X, Wang D, Gao Z, Wang C, Yao C, Ma Y (2011) Textured $Sr_{1-x}K_xFe_2As_2$ superconducting tapes with high critical current density. Physica C 471:1689
13. Gao Z, Wang L, Yao C, Qi Y, Wang C, Zhang X, Wang D, Wang C, Ma Y (2011) High transport critical current densities in textured Fe-sheathed $Sr_{1-x}K_xFe_2As_2$ superconducting tapes. Appl Phys Lett 99:242506
14. Gao Z, Ma Y, Yao C, Zhang X, Wang C, Wang D, Awaji S, Watanabe K (2012) High critical current density and low anisotropy in textured $Sr_{1-x}K_xFe_2As_2$ tapes for high field applications. Sci Rep 2:998
15. Parrell JA, Dorris SE, Larbalestier DC (1994) On the oxide core density, filament uniformity, and critical current density of $(Bi, Pb)_2Sr_2Ca_2Cu_3O_x$ tapes. Adv. Cryo. Eng. 40:193
16. Kovac P, Husek I, Pachla W (1997) Ceramic core density and homogeneity in BSCCO/Ag tapes. IEEE Trans Appl Supercond 7:2098
17. Grasso G, Jeremie A, Flukiger R (1995) Optimization of the preparation parameters of monofilamentary Bi(2223) tapes and the effect of the rolling pressure on J_c. Supercond Sci Technol 8:827
18. Pcahla W, Marciniak H, Szulc A, Wroblewski M, Kovac P, Husek I, Melisek T (1997) Investigation of texture formation and phase transition in press-, CIP- and roll-sintered Ag-sheathed Bi(2223) tapes. IEEE Trans Appl Supercond 7:2090
19. Grasso G, Perin A, Hensel B, Flukiger R (1993) Pressed and cold rolled Ag-sheathed Bi(2223) tapes: a comparison. Physica C 217:335
20. Han Z, Skov-Hansen P, Freltoft T (1997) The mechanical deformation of superconducting BiSrCaCuO/Ag composites. Supercond Sci Technol 10:371
21. Yao C, Lin H, Zhang X, Wang D, Zhang Q, Ma Y, Awaji S, Watanabe K (2013) Microstructure and transport critical current in $Sr_{0.6}K_{0.4}Fe_2As_2$ superconducting tapes prepared by cold pressing. Supercond Sci Technol 26:075003
22. Lotgering FK (1959) Topotactical reactions with ferrimagnetic oxides having hexagonal crystal structures-I. J Inorg Nucl Chem 9:113
23. Yao C, Lin H, Zhang X, Dong C, Wang D, Zhang Q, Ma Y, Awaji S, Watanabe K (2015) Transport critical current density of $Sr_{0.6}K_{0.4}Fe_2As_2$/Ag superconducting tapes processed by flat rolling and uniaxial pressing. IEEE Trans Appl Supercond 25:7300204

24. Togano K, Gao Z, Taira H, Ishida S, Iyo A, Kihou K, Eisaki H, Matsumoto A, Kumakura H (2013) Enhanced high-field transport critical current densities observed for ex situ PIT processed Ag/(Ba, K)Fe_2As_2 thin tapes. Supercond Sci Technol 26:065003
25. Togano K, Matsumoto A, Kumakura H (2012) Fabrication and transport properties of ex-situ powder-in-tube (PIT) processed (Ba, K)F_e2A_s2 superconducting wire. Solid State Commun 152:740
26. Zeng R, Ye B, Horvat J, Guo YC, Zeimetz B, Yang XF, Beales TP, Liu HK, Dou SX (1998) Critical current density significantly enhanced by hot pressing in Bi-2223/Ag multifilamentary tapes. Supercond Sci Technol 11:1101
27. Sato K, Kobayashi S, Nakashima T (2012) Present status and future perspective of bismuth-based high-temperature superconducting wires realizing application systems. Jpn J Appl Phys 51:010006
28. Gao Z, Togano K, Matsumoto A, Kumakura H (2014) Achievement of practical level critical current densities in $Ba_{1-x}K_xFe_2As_2$/Ag tapes by conventional cold mechanical deformation. Sci Rep 4:04065
29. Zhang X, Yao C, Lin H, Cai Y, Chen Z, Li J, Dong C, Zhang Q, Wang D, Ma Y, Oguro H, Awaji S, Watanabe K, Realization of practical level current densities in $Sr_{0.6}K_{0.4}Fe_2As_2$ tape conductors for high-field applications. Appl Phys Lett 104:202601

Chapter 5
Mechanism of J_c Enhancement for 122-Type IBS Tapes

For the iron-based superconducting wires and tapes, their transport critical current density (J_c) is calculated by dividing the critical current (I_c) by the cross sectional area of the superconducting cores. Owing to the strong intrinsic flux pinning of iron-based superconductors (IBS), at present the transport J_c in IBS wires and tapes made by PIT method is mainly limited by weak grain connectivity and high-angle grain boundaries. As mentioned above, some extrinsic material defects induced during the wire fabrication processes, including the oxidization and impurity of precursor, the low density stemmed from powder packing process, the grain crush, microstructural cracks and residual stress caused by mechanical deformation, and composition segregation, residual pores and secondary phase formed in the heat treatment process, will degrade the grain connectivity. The grain connectivity was usually characterized by observing the micro morphology with SEM. In this chapter, we use magneto-optical imaging (MOI) technique to characterize the granularity, homogeneity, and quantitatively inter-granular J_c for the tape samples. The investigation of MOI for the IBS samples is more macroscopic than SEM, and more microscopic than transport I_c measurement, so it can help to build up a connection between the microstructure of grains and the J_c performance of wire and tape samples. On the other hand, the misoriented grains in IBS, especially high-angle grains, will result in weak-link effect at grain boundaries. The weak-linked grain boundaries will exponentially suppress the inter-grain currents when the misorientation angle is larger than a certain value (9° for 122-type IBS) [1, 2], no matter how tight the grains bond with each other. A common indication for the weak-link behavior is a hysteretic effect of transport J_c measured in increasing and decreasing fields in IBS wires and tapes [3–5]. The weak-linked grain boundaries can be indirectly indexed by the degree of grain texture with XRD analysis or by qualitative SEM observation. With high resolution transmission electron microscopy (HRTEM), we will characterize and analyze the weak-linked grain boundaries on the atomic scale for IBS tapes. For high-field magnet applications, large current carrying capability, strong pinning characteristics and low superconductivity anisotropy are desirable. Since in the high-field regime

C. Yao, *Fabrication and Properties of High-Performance 122-Type Iron-Based Superconducting Wires and Tapes*, Springer Theses,
https://doi.org/10.1007/978-981-10-5184-5_5

J_c is mainly limited by flux motion within the IBS grains, strong flux pinning ability is also essential for superior J_c-field performance. In additon to the grain coupling analysis, we will also analyzed the flux pinning behavior of high-performance IBS tape samples, which shows the mechanism for the great current carrying ability of IBS tapes in high fields.

5.1 Grain Coupling Analysis

5.1.1 *Magneto-Optical Imaging Characterization*

Magneto-optical imaging (MOI) is an effective technique to visualize the local and real-time distribution of the magnetic flux. In order to improve the transport properties of the superconducting wires and tapes, it is necessary to investigate the intergranular J_c and the magnetic flux distribution of the superconducting cores. The MOI characterization has been widely used for cuprate superconductors [6–10], and after the discovery of iron-based superconductors, it has been used to investigate the superconducting properties for 1111-, 122- and 11-type IBS single crystals and polycrystals [11–15]. Recently, MOI characterization was employed to reveal the detailed information such as the granularity, homogeneity, intra- and intergranular J_c for Ba-122 and Sm-1111 wires and tapes [16–19]. As mentioned above, the grain connectivity is crucial to achieve high transport J_c for superconducting wires and tapes. In Chaps. 3 and 4, the grain connectivity for Ba-122 and Sr-122 IBS tapes is analyzed using SEM. However, with the results obtained by SEM, we can only discuss the grain connectivity with respect to the microstructure, which cannot give a quantitative and more accurate evaluation for the local current carrying ability inside the samples. Therefore, MOI characterization was carried out on Fe-sheathed Sr-122 tapes made by flat rolling and heat treated at 850–900 °C, and their J_c performance at various temperatures was studied and compared with the results obtained by standard four-probe transport measurement and magnetic measurement.

The MO images of the magnetic flux penetration after zero-field cooling (ZFC) the Sr-122 tape samples to 5 K are presented in Fig. 5.1. The external magnetic fields were applied perpendicular to the sample surface. The bright regions represent the areas where the magnetic flux exists. It can be observed that due to the strong shielding currents, the magnetic flux does not penetrate into the whole sample when the external field is lower than 20 mT. When the field is further increased, it gradually moves towards the sample center from the edge, but even when the applied is increased to 50 mT, it still just partially penetrates the sample. This result indicates strong bulk circulating currents in the sample, which can be ascribed to the good grain connectivity and the reduced mismatch of the grains with textured microstructure. The magnetic flux penetration patterns of our tapes are similar to that of the bulk material synthesized by the HIP technique [17], and completely different from that of the weak linked 122 IBS wire sample [16].

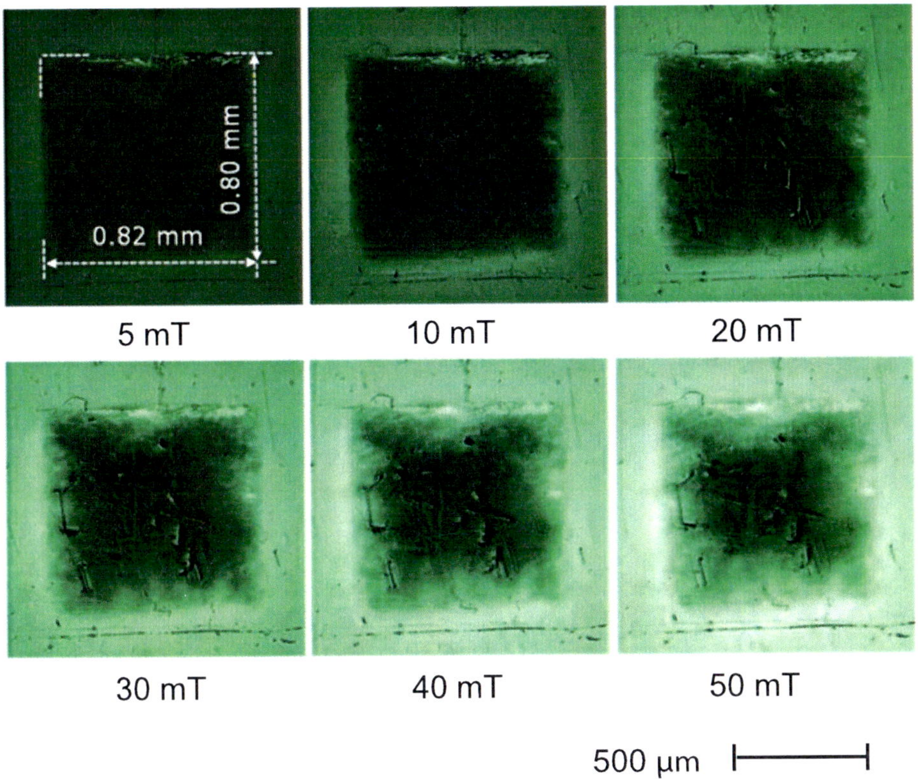

Fig. 5.1 Magneto-optical images in Meissner state showing magnetic flux penetration into the Sr-122 samples in external magnetic fields of 5, 10, 20, 30, 40 and 50 mT after zero-field-cooling (ZFC) the samples to 5 K. The applied field is perpendicular to the sample plane. The dimensions of the samples are indicated with white dashed lines. Reprinted from Yao et al. [20]

In order to reveal the relationship between the global J_c and the microstructure of the Sr-122 tape samples, the distribution of remnant trapped fields for the samples was measured after they were cooled to various temperatures below the T_c, applied with an external field, and then remove the applied field. As shown in Fig. 5.2, the Sr-122 tape samples were first zero-field-cooled to 5, 10, 15, 20, 25 and 30 K from the normal state (i.e., temperature above the T_c), respectively, then an external field of 400 Oe was applied for 0.2 s. After removing the applied field, the remanent trapped fields in the samples can be observed. When the temperature was lower than 20 K, the magnetic flux still does not reach the center of the Sr-122 sample, indicating very strong shielding currents in the sample. When the samples was cooled to 25 and 30 K, which is near the T_c, the magnetic flux reaches the central part of the samples, and a very uniform roof top pattern of the remanent trapped fields can be clearly observed, testifying the almost uniform bulk-scale current flow in the samples. This result indicates that the samples consist of very homogeneous Sr-122 phases, since even at the temperatures near the T_c, the samples shows good uniformity for the current carrying ability.

Fig. 5.2 Magneto-optical (MO) images in remnant state showing trapped magnetic flux in the zero-field-cooled (ZFC) Sr-122 samples after cycling the applied field up to 40 mT at 5, 10, 15, 20, 25 and 30 K. The applied field is perpendicular to the sample plane

When we increased cycling field from 400 to 800 Oe, the magnetic flux still does not reach the center of the Sr-122 sample at 5 K, but at 20 K, it reach the central part of the samples, as presented in Fig. 5.3a, b. In these figures, the magnetic induction line profiles along the red line on the MO image were plotted. For the MO image at 5 K, the intergranular J_c can be estimated using the equation [13, 21],

$$J_c = \frac{\mathrm{c}}{4d} \frac{H_{\mathrm{ex}}}{\cosh^{-1}[w/(w-2p)]} \tag{5.1}$$

where H_{ex} is the external field, d is the sample thickness, w is the sample width, and p is the flux penetration distance from the sample edge. In Fig. 5.3a the right blue dashed line indicates the edge of the sample, and the left blue dashed line indicates the front edge of the flux penetration. Therefore, with $H_{\mathrm{ex}} = 800$ Oe, $c = 10$, $d =$

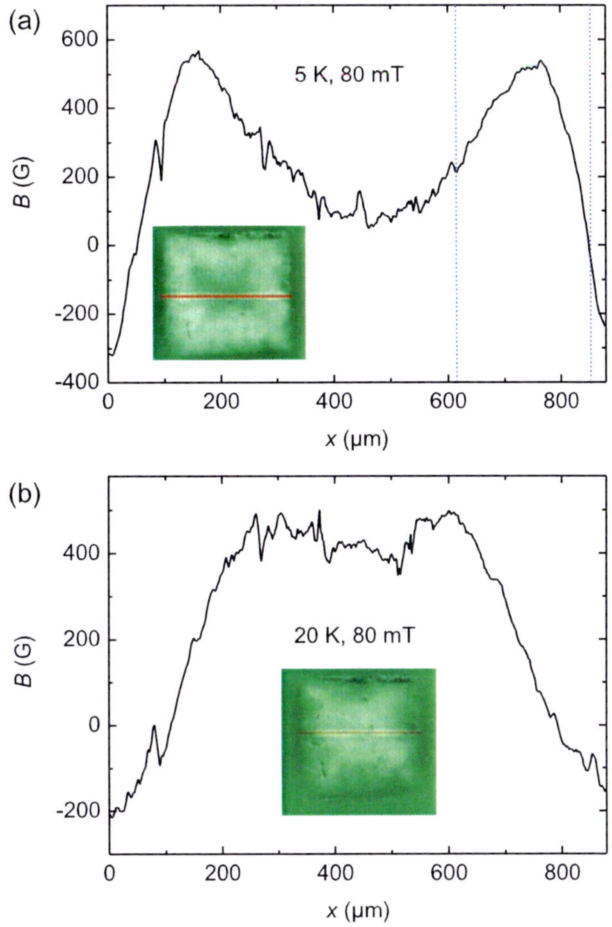

Fig. 5.3 Magneto-optical (MO) images in the remnant state showing trapped magnetic flux and the magnetic induction line profiles along the red lines for the Sr-122 zero-field-cooled (ZFC) samples. **a** Sr-122 sample after cycling the applied field up to 80 mT at 5 K. The blue dashed lines indicate the distance between the sample edge and the front edge of the magnetic flux penetration towards the sample center. **b** Sr-122 sample after cycling the applied field to 80 mT at 20 K. The applied field is perpendicular to the sample plane. Reproduced from Yao et al. [20]

100 μm, $w = 820$ μm and $p = 274$ μm, the J_c at 5 K for the Sr-122 sample is calculated to be about 1.1×10^5 A/cm^2.

In Fig. 5.3b, at 20 K we obtain a very uniform roof top pattern of the fully trapped magnetic flux, and the line profile calculated from the MO image data also shows a roof shape, suggesting bulk currents flow over the whole sample. In this case, for a thin disk-shaped superconductor in the remnant state with trapped field perpendicular to the plane, the critical state occurs through the thickness [22–24], so we can roughly estimate the intergranular J_c using the equation [14, 21],

$$J_c = \frac{\Delta B}{d} \tag{5.2}$$

where ΔB is the trapped field, and d is the thickness of the sample. According to the magnetic induction line profile, the global J_c at 20 K of this sample is estimated up to

4×10^4 Acm^{-2}. Similarly, the intergranular J_c at 10, 15, and 25 K can be calculated with these two equations, respectively.

To further illustrate the effect of grain connectivity on the J_c of the IBS tapes, we compared the MOI results of the Ba-122 round wire with Ag addition [16] with that of our Sr-122 tapes with Sn addition. The remanent state of the Ba-122 wire sample at 5 K shows a very non-uniform magnetic flux distribution, and the trapped magnetic flux is distributed in some local regions with brighter magneto-optics, indicating that the intergranular current in the sample is much smaller than the intragranular current, which is quite different from the results of our Sr-122 tape samples at 5 K shown in Figs. 5.2 and 5.3, where the induced current flows uniformly throughout the whole samples. The distribution of the intensity of the induced magnetic field along the horizontal line for the Ba-122 wire sample also shows large fluctuations. For the Meissner state of the sample at 5 K, the Ba-122 wire sample has been completely penetrated by an applied magnetic field of 20 Oe, and only some small local regions are still in the Meissner state, which is different from the case shown in Fig. 5.1, where our Sr-122 tape sample can completely shield the external magnetic field of 200 Oe. These results all demonstrate that the grain connectivity of our Sr-122 tape samples is much better than that of the Ba-122 wire samples. As a result, though the field dependence of the transport J_c is similar between Sr-122 tape samples and the Ba-122 wire samples, the J_c values of the former is one order of magnitude higher than that of the latter.

To confirm the intergranular J_c estimated from the MOI results, we measure the magnetic hysteresis at various temperatures of the Sr-122 tape samples used in the MOI measurements, as shown in Fig. 5.4a. The magnetic fields were applied perpendicular to the sample surface. The bulk magnetic J_c can be derived from the magnetic hysteresis loops using the Bean model [25],

$$J_c = \frac{20\Delta M}{a(1 - b/3a)} \tag{5.3}$$

where ΔM is the difference between the magnetization when sweeping fields up and down, a and b are the sample widths ($a < b$).

Figure 5.4b shows J_c for the Sr-122 tape samples as a function of magnetic field at various measuring temperatures from 5 to 30 K, as calculated from magnetization loops M-H using the critical state Bean model. It can be noted that the magnetic J_c of the samples exceeds 10^5 A/cm^2 at both 5 and 10 K in self-field, and even at a temperature of 30 K, the J_c value is close to 10^4 A/cm^2, indicating that our sample has good uniformity and crystallinity. On the other hand, at the temperature of 15 K, the J_c decreased by about one order of magnitude when the applied magnetic field increased from 0 to 3 T; while at the temperature of 25–30 K, when the applied field increased from 0 to 3 T, the J_c decreases by about two orders of magnitude. Therefore, the field dependence for the magnetic J_c is stronger at higher temperatures. In order to compare the results obtained by MOI and magnetization loops, the temperature dependences of the intergranular J_c estimated from the M-H and MOI measurements

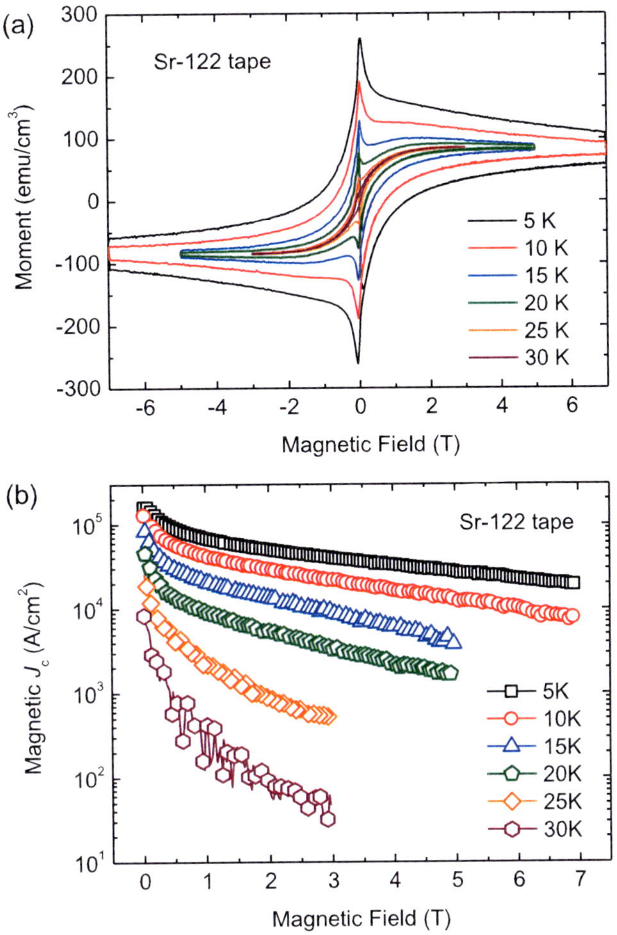

Fig. 5.4 **a** The magnetic hysteresis loops and **b** the field dependence of magnetic J_c for the Sr-122 tapes at various temperatures. The applied field is perpendicular to the sample plane. Reprinted from Yao et al. [20]

are plotted in Fig. 5.6. Generally the results obtained by *M-H* measurements are in accordance with that obtained by MOI measurements. Therefore, as a visualized characterization technique, the MOI measurement is also accurate for J_c evaluation.

The magnetic field dependence of the transport J_c obtained by a standard four-probe method at 4.2, 15, 20 and 25 K for the Sr-122 tape is presented in Fig. 5.5. Similar to the result of magnetic J_c, the transport J_c also exhibits stronger field dependence at higher temperatures. Nevertheless, at 20 K, the transport J_c reached ~10^4 A/cm^2 in self-field and ~650 A/cm^2 at 10 T, respectively. Even measuring at 25 K, we observed J_c values of over 2.1×10^3 A/cm^2 in self-field and 160 A/cm^2 at 5 T, respectively. These promising results at high temperatures suggest that the Sr-122 superconducting tapes have a strong potential in high-field magnets operating at temperature range around 20 K, which can be attained by liquid hydrogen or cryogenic cooling.

We compare the critical current densities at different temperatures obtained by the four-lead method with magnetization measurements, as shown in Fig. 5.6. The

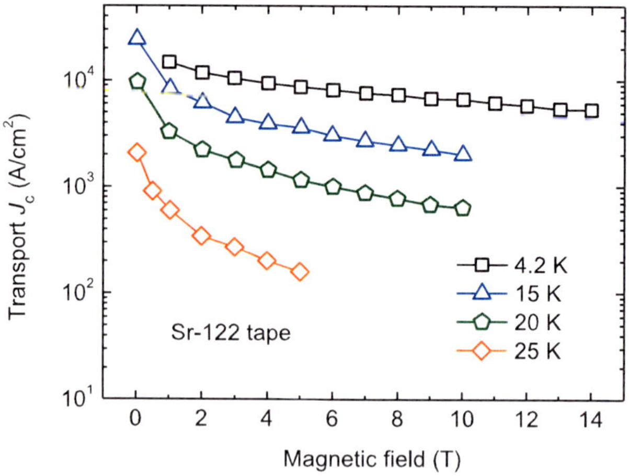

Fig. 5.5 Field dependence of the transport J_c at various measuring temperatures for the Sr-122 tapes. The applied field is parallel to the tape plane

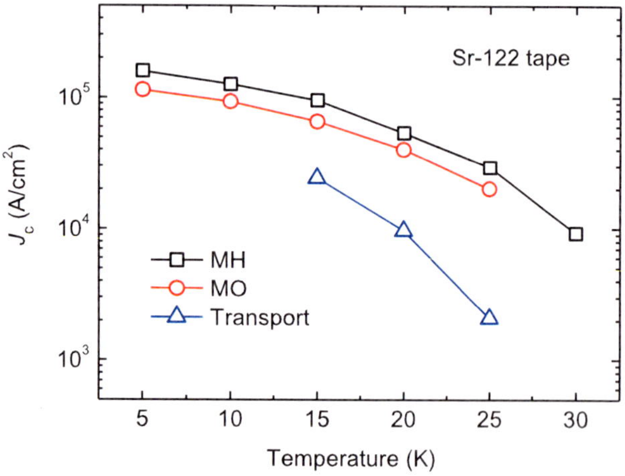

Fig. 5.6 Temperature dependence of the magnetic J_c estimated from the *M-H* and MOI measurements and the transport J_c measured by using the four-probe method at various temperatures for Sr-122 tapes

magnetic J_c estimated from the *M-H* measurement is higher than the transport J_c, and it seems that at higher temperatures, the difference between them is greater. At 20 K, the magnetic J_c is 5.4×10^4 A/cm^2, while the transport J_c is 1.0×10^4 A/cm^2. This is partly due to the fact that the results obtained by the magnetic measurement method contain the contribution of the intragranular current in the sample, especially in the low-temperature region. Another possible reason is that the transport critical current can be affected by inhomogeneities along the length of the Sr-122 tapes, since the samples for the transport measurement are much larger than that used for *M-H* and MOI measurements. Therefore, the current-carrying performance of our prepared Sr-122 iron-based superconducting tape still has room for further improvement.

For the development of high-performance iron-based superconducting tapes, one of the most important issues is improving the grain connectivity. As demonstrated in Chap. 3, doping of silver and lead can fill the voids between the grains, or acts as a growth flux for IBS grains during heat treatment. In Chap. 4, cold deformation processes such as rolling and pressing can enhance the mass density of the superconducting phase, so as to improve the grain connectivity. Furthermore, heat treatment that helps to eliminate micro cracks induced by mechanical deformation is also indispensable. After the grain connectivity is remarkably improved, we can further enhance the J_c performance by optimizing the grain orientation for iron-based superconducting tapes.

5.1.2 High-Resolution TEM Characterization

As presented in Chap. 4, with employing cold pressing process, the grain connectivity and grain alignment can be effectively improved, resulting in remarkable enhancement for the J_c performance of the Sr-122 tapes. However, some residual transverse cracks induced during flat rolling process may still exist after cold press process and the final heat treatment. Then we introduce hot pressing technique, which combines the uniaxial press deformation with the heat treatment in one procedure, to further improving the current carrying ability for the Sr-122 tapes. As a result, a high J_c reaching the practical level of 10^5 A/cm^2 at 4.2 K and 10 T was achieved for the first time in the Sr-122 IBS tapes processed with hot press technique. In order to have a detailed understanding of the structure, composition and grain coupling at grain boundaries, high-resolution TEM characterization is needed to investigate the mechanism for enhancing intergranular J_c.

The coherence length of the 122-type iron-based superconductor is relatively small with $\xi(0\text{ K})$ less than 2 nm and $\xi(4\text{ K}) \sim 2.6$ nm. High-energy grain boundaries whose size exceeds the coherence length will lead to the decay of electronic coupling between grains, which suppress the superconducting current. As shown in Fig. 5.7a, for the Sr-122 polycrystalline samples with low intergranular J_c of about 10^3 A/cm^2

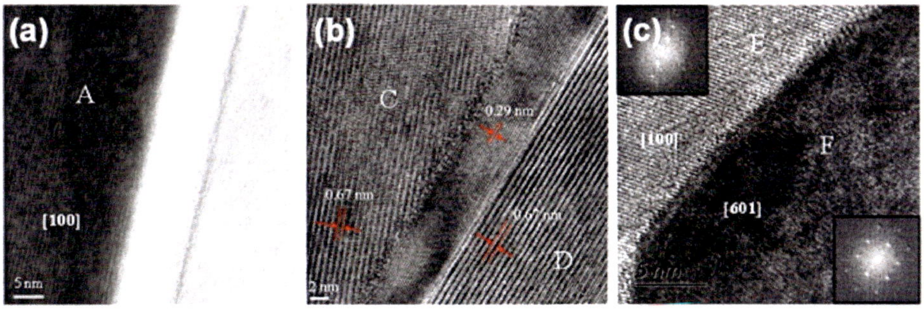

Fig. 5.7 High-resolution STEM image of the grain boundaries of Sr-122 tapes. **a** A grain boundary containing an amorphous layer about 10 nm in thickness [26]. **b** A grain boundary containing a Sn layer about 3–4 nm in thickness [27]. **c** A clean grain boundary showing lattice fringes of the two grains meeting without an amorphous contrast or secondary phase (this work)

at 4.2 K in self-field, it is found that between two Sr-122 grains A and B, there is an amorphous layer of about 10 nm in thickness, which is larger than the coherence length of this superconductor. Furthermore, as analyzed by electron energy loss spectroscopy (EELS) at atomic resolution in a scanning transmission electron microscope (STEM), there is significant oxygen enrichment in the amorphous layer and crystallites. Such boundary features indicate that a large fraction of grain boundaries act as critical current barriers, and that reducing of the oxygen content during the fabrication process will be essential for attaining high intergranular J_c [26]. In Fig. 5.7b, for the grains C and D in Sr-122 tapes with transport J_c as high as 2.3 × 10^4 A/cm^2 (4.2 K, 10 T), a Sn layer about 3–4 nm thick was found between them [27]. As a metal additive, Sn can improve the metallic character of secondary phases at GBs, bridging the pores and cracks to enhance the grain connectivity, and also help to prevent the formation of glassy phases and amorphous layers at GBs. However, if the thickness of the Sn layer is larger than the coherence length, as showing in this figure, so it will suppress the intergranular current. Another positive effect brought by added Sn is to improve homogenization of the chemical composition of the Sr-122 phase, since it can help to eliminate the FeAs wetting phase at the grain boundaries. For the hot pressed Sr-122 tapes with high transport J_c of 10^5 A/cm^2 at 4.2 K and 10 T, we can observe lots of clean grain boundaries, as showing in Fig. 5.7c that lattice fringes of the two grains E and F meeting without an amorphous contrast or secondary phase. Further, for these clean grain boundaries, we should pay attention on the misorientation angles of grains, since the presence of the weak-links at high-angle grain boundaries will lead to exponentially decreases for the transport J_c.

It is well known that the inter-grain currents in HTS including iron-based superconductors and cuprate superconductors are intrinsically deteriorated by the high-angle grain boundaries (GBs). The transport properties of bicrystal GBs with $\theta_{GB} = 3 \sim 45°$ was investigated using Co-doped Ba-122 epitaxial thin films grown on MgO and LSAT bicrystal substrates, as illustrated in Fig. 5.8a [28]. The θ_{GB} dependence of J_c^{BGB} normalized by the film J_c at 4 K for lower-angle BGBs was plotted in Fig. 5.8c. Compared with the corresponding data of YBCO BGBs, the exponential decay starts at a critical angle θ_c of approximately 9–10°, substantially larger than the 3–5° reported for YBCO BGBs, and is more gradual when the θ_{GB} is larger than the critical angle. Figure 5.8b shows [001] plan-view high-resolution TEM images of the Co-doped Ba-122 BGB junctions with various GB angles. It can be seen that periodic misfit dislocations are clearly along the BGBs with intervals of ~ 5.0 nm for $\theta_{GB} = 4°$ and 1.2 nm for $\theta_{GB} = 24°$. Obviously, the GB dislocation spacing for the BGB junctions with $\theta_{GB} = 24°$ is larger than the coherence length ξ(4 K) of 2.6 nm, and in this case the supercurrent cannot pass through the BGBs, giving rise to the weak-link behavior. For higher-angle BGBs, the distance between the dislocations becomes smaller and eventually dislocations overlap each other. For instance, at the BGB junctions with $\theta_{GB} = 45°$, due to the GB dislocation spacing is close to the lattice parameter, we observe no periodic misfit dislocations but blurred lattice fringes across the entire region. In addition, there would be other factors that affect the GB transport properties. For instance, the dislocation cores formed along BGBs

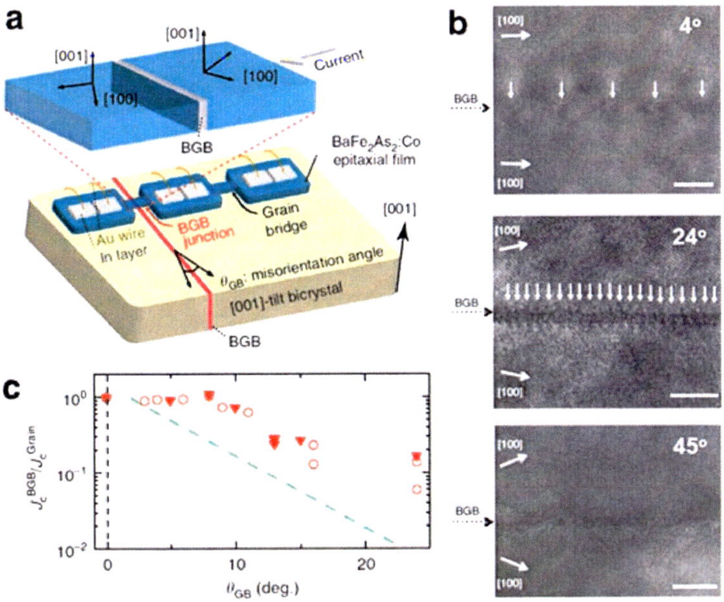

Fig. 5.8 Critical current density (J_c) versus misorientation angle (θ_{GB}) and [001] plan-view HR-TEM images of the Co-doped Ba-122 bicrystal grain boundary (BGB) junctions on MgO bicrystal substrates. **a** Device structure of the BGB junctions and grain bridges. The upper rectangular solid is an enlargement at the BGB junction. **b** BGB junctions with misorientation angle θ_{GB} of 4°, 24°, and 45°, respectively. The directions of BGB junctions are indicated by dashed arrows and the [100]-axes are symmetrically tilted from the BGB lines. The misfit dislocations are marked by the down-pointing arrows. Each horizontal bar indicates 5 nm scale. **c** The ratio of the intragrain J_c (J_c^{Grain}) and J_c^{BGB} to $\theta_{\mathrm{GB}} = 0° \sim 25°$ at 4 K. Open and closed symbols show the ratios of samples on MgO and LSAT bicrystals, respectively. The dashed green line shows the result of the YBCO BGB junctions. Reproduced from Katase et al. [28]

can produce residual strains, which induces a local transition to an antiferromagnetic phase and forms insulating regions near dislocation cores in cuprate superconductors [29].

For iron-based superconducting wires and tapes, the grain connections GB is more complex than that for bicrystal films. Similar to the case in Bi-2223 tapes [31, 32], ignoring the grain twisting around the *c*-axis or *a*-axis with respect to each other for simplicity, the connections of Sr-122 grains piled up along the *c*-axis can be recognized as two types: I. platelet-stacked (Grains 1 & 2) and II. edge-bonded (Grains 2 & 3), and the GBs of these two types in the case of imperfect *c*-axis texture are indicated by red arrows in Fig. 5.9a. When all the Sr-122 grains are perfectly oriented along the *c*-axis, the supercurrents (marked with blue arrows) are free to flow across edge-bonded GBs along the *ab* planes. If the current path in this direction is blocked by extrinsic defects such as microcracks, porosity and non-superconducting phases, the transport currents will transfer along paths in the *c*-axis direction, and flow to the adjacent basal grain by crossing platelet-stacked GBs, as described in Bi-based cuprates [33, 34]. However, if the grains are not perfectly

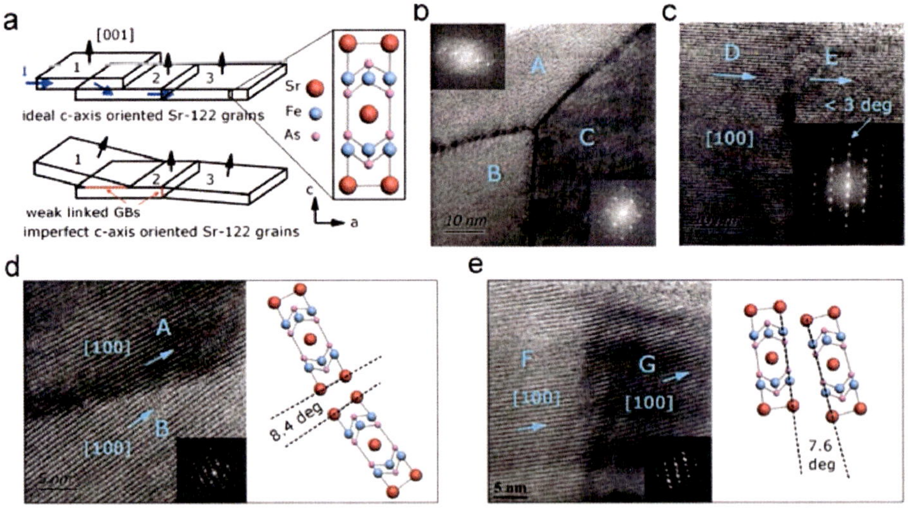

Fig. 5.9 HRTEM study on the grain boundaries (GBs) of hot pressed Sr-122 tapes. **a** Schematic diagram of *c*-axis oriented Sr-122 grains inside tapes in the longitudinal-sectional view, and the sections of the atomic structure of Sr-122 in projection along the [010] direction. **b** HRTEM observation of the Sr-122 superconducting core in longitudinal direction of the tape, showing clean GBs of three grains (marked as A, B and C). **c** HRTEM image showing a very small misorientaion angle <3° of two grains (marked as D and E). **d** Enlarged image of the GB along *ab*-plane between grain A and B, showing a low *c*-axis tilt misorientaion angle of 8.4°. **e** Enlarged image of the GB between grain F and G along *bc*-plane, showing a low *c*-axis tilt misorientaion angle of 7.6°. The insets in (**b**), (**c**), and (**e**) are the fast fourier transformation of the HRTEM images. The inset in (**d**) is the selected-area electron diffraction image of the grains. Reproduced from Zhang et al. [30]

textured in the *ab*-plane, i.e. their *c*-axis are tilted by an angle with respect to each other, the misoriented crystal lattice will meet at the platelet-stacked and edge-bonded GBs, so the supercurrents flow across them will be decreased due to the weak-link effect. Fortunately, a high transport supercurrent does not require perfect *c*-axis texture, because the misorientation with low-angle does not seriously deteriorate the supercurrent. According to the study for [001]-tilt $Ba(Fe_{1-x}Co_x)_2As_2$ bicrystals [28, 35], there is almost no weak-link behavior at GBs with misorientation angle $\theta_{GB} \leq 9°$, and the intergranular J_c is similar to intragranular J_c at this circumstance. Hence, high transport J_c can be obtained for 122 iron-pnictide tape with low-angle GBs.

Figure 5.9b shows a typical image of high resolution transmission electron microscopy (HRTEM) image consists of three Sr-122 grains (marked as A, B and C) in longitudinal view of the hot pressed tapes, in which the sample was tilted so that the grain boundary was almost parallel to the incident electron beam. It can be seen that the GBs between them are very clean, and no amorphous contrast is observed, indicating the good Sr-122 phase purity and the well-connected Sr-122 grains to each other. However, as mentioned above, due to the weak-link effect at the high-angle GBs, only good grain connectivity is not sufficient to guarantee high transport supercurrents. Therefore, we must check the grain misorientations at GBs. Figure 5.9c presents two Sr-122 grains (marked as D and E) with a small misorientation angle

<3°, which can clearly be seen in the fast fourier transformation of the HRTEM image in the inset. Enlarged image of the GB between grain A and B is shown in Fig. 5.9d, selected area electron diffraction (SAED) shown in the inset confirmed that both the lattice fringe in A and B grains refer to the [100] direction. The GB between grain A and B belongs to the platelet stacked GB mentioned above, and the [100] tilt misorientaion angle is 8.4°, Fig. 5.9e shows the GB between other two grains (marked as F and G), which is in accordance with the edge-bond case, with a small misorientaion angle of 7.6°. Besides, lots of low-angle GBs can be observed in the pressed samples, so the transport J_c will not suffer much depression when it run across such GBs with small misoriented angles <9°. Therefore, similar to the railway-switch mechanism in the *c*-axis textured Bi-2223 tapes, platelet-stacked and edge-bonded Sr-122 GBs with low misorientation angles can also ensure enough current path for high supercurrent.

In order to further investigate the microstructure of the hot pressed Sr-122 tapes, the superconducting core processed by focused ion beam (FIB) technique was submitted to TEM examination for the element distribution of Sr, K, Fe and As, respectively. As shown in Fig. 5.10, the good homogeneity of our tapes has been further confirmed by energy dispersive X-ray spectroscopy (EDS) mapping based on TEM, which demonstrates that the element distribution of Sr-122 phase is homogeneously distributed in the superconducting core. Therefore, there is no secondary phase in the observed core region, which is in accordance with the clean GBs shown in Fig. 5.9. The segregation of secondary phases (such as oxygen-rich amorphous layers) around GBs of the pnictide superconductor is frequently cited as the limiting factor for J_c because of the concomitant degradation in intergranular connectivity. Obviously, densified superconducting core, elimination of the cracks, and clean grain boundaries attribute to the good grain connectivity in the hot pressed Sr-122 tapes.

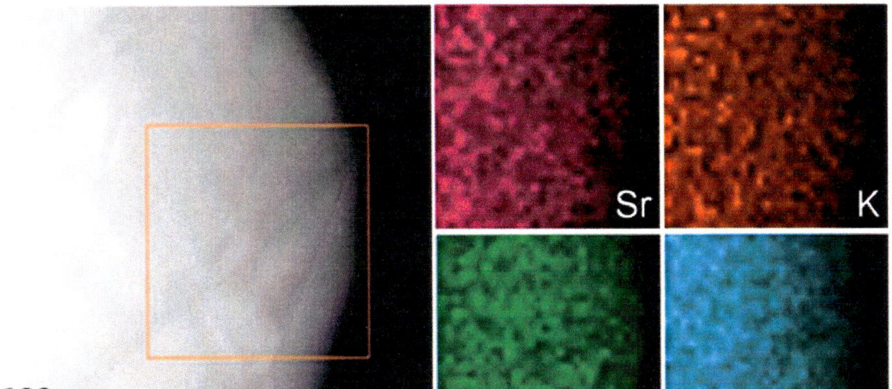

Fig. 5.10 TEM image and EDS mapping images of the core region of hot pressed Sr-122 tapes. The elements Sr, K, Fe and As of superconducting phase were homogeneously distributed in the superconducting core. Reproduced from Zhang et al. [30]

Based on such improved grain connectivity, the grain alignment also plays an important role in the contribution to the percolation behavior of the current path in the Sr-122 tapes. The c-axis textured grains result in a high fraction of low-angle GBs as observed by high-resolution TEM.

5.2 Flux Pinning Analysis

5.2.1 *Upper Critical Field and Irreversible Field*

Well-coupled grains in iron-based superconducting wires can only alleviate the suppression of inter-grain J_c by weak-linked grain boundaries, and the main factor that determines the field dependence of J_c is the flux pinning. Therefore, it is also important to study the pinning properties for the iron-based superconducting wires. In order to conduct a more comprehensive study for the flux pinning properties, two kinds of Ag-sheathed Sr-122 IBS tapes were analyzed, one is flat rolled tapes, and the other is hot pressed tapes. The fabrication process for them is similar to that described in Sects. 4.3 and 4.4, respectively.

Figure 5.11a, b show the temperature dependence of resistivity at various magnetic fields for the flat rolled and hot pressed Sr-122 tapes, respectively. The applied field is perpendicular to the tape plane. The T_c^{onset} and T_c^{zero} at 0 T is about 34.8 and 33.9 K for the flat rolled tapes, and 35.4 and 34.7 K for the hot pressed tapes. The transition width $\Delta T = T_c^{\mathrm{onset}} - T_c^{\mathrm{zero}}$ is 0.9 K for the flat rolled tapes and 0.7 K for the hot pressed tapes, respectively, indicating good homogeneity of the superconducting cores for these two kinds of tapes. In addtion, the both T_c^{onset} and T_c^{zero} of hot pressed tapes are about 0.6–0.8 K higher than that of flat rolled tapes, indicating higher phase quality can be obtained by hot pressing. When the applied field increases, the T_c^{onset} and T_c^{zero} change towards lower temperatures for all the tapes, and the values drop more dramatically for T_c^{zero}. In addtion, it is obvious that the decrease of transtion temperatures with the increase of field is more dramatic in flat rolled tapes than that in hot pressed tapes, which is ascribed to weak-coupled grains for the former.

Taking a criterion of 90 and 10% of normal state resistivity on the resistive transition curves in Fig. 5.11a, b, the upper critical field H_{c2} and irreversible field H_{irr} for the flat rolled and hot pressed tapes at various temperatures are estimated, as plotted in Fig. 5.12. Similar to the behavior of low-T_c superconductors [36], the in-field resistive transition of the pressed Sr-122 tapes exhibits nearly no broadening, resulting in relatively closer $H_{c2}(T)$ and $H_{\mathrm{irr}}(T)$ curves. The upper critical field at zero-temperature $H_{c2}(0)$ can be calculated by using the Werthamer-Helfand-Hohenberg (WHH) formula:

$$H_{c2}(0) = -0.693 T_c \frac{\mathrm{d}H_{c2}}{\mathrm{d}T} \tag{5.4}$$

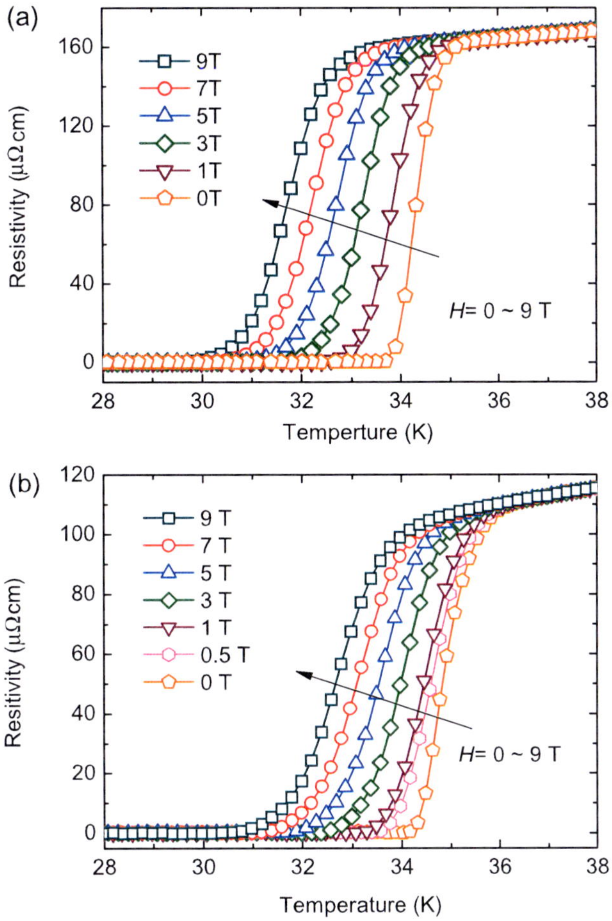

Fig. 5.11 Temperature dependence of resistivity at various magnetic fields for **a** flat rolled tapes and **b** hot pressed Sr-122 tapes. The applied field is perpendicular to the tape plane

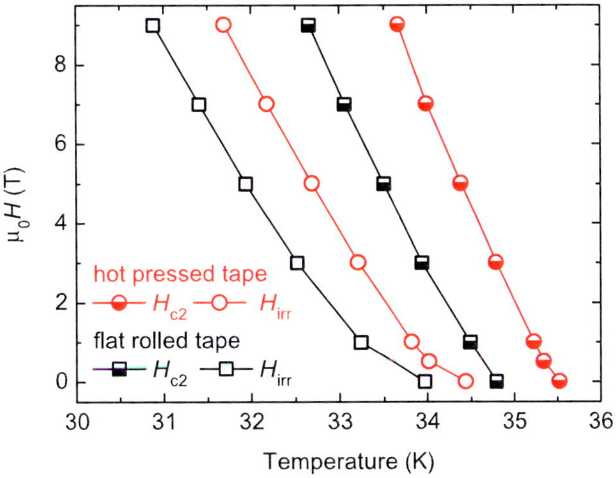

Fig. 5.12 Upper critical field $\mu_0 H_{c2}(T)$ and irreversibility field $\mu_0 H_{irr}(T)$ for the flat rolled tapes and hot pressed Sr-122 tapes at various temperatures. The applied field is perpendicular to the tape plane

The $B_{c2}(0)$ for the flat rolled and hot pressed tapes are estimated to be 110 and 119 T, respectively. These results are comparable to that measured on the K-doped Ba-122 single crystal [37]. Moreover, the Ginzburg-Landau coherence length ξ_{GL} can be determined by the fomula:

$$\xi_{GL} = \sqrt{\frac{\Phi_0}{2\pi\mu_0 H_{c2}}} \tag{5.5}$$

where $\Phi_0 = 2.07 \times 10^{-7}$ Oe cm^2, so the calculated $\xi_{GL}(0) = 1.90$ and 1.98 nm for flat rolled and hot pressed samples. In addition, it should be noted that the H_{irr} determines the regime in the *T-H* phase diagram where this material can be operated. Therefore, the large H_{irr} exhibited in the flat rolled and hot pressed samples signifies promising future of IBS material for high-field applications.

5.2.2 *Pinning Energy*

Thermally activated flux flow is manifested in the broadening of superconducting transitions under various magnetic fields. Since the resistance in the low-resistance region is induced by the creep of vortices [38], according to the thermally activated flux flow (TAFF) model, the temperature dependences of the resistivity ρ can be described by the equation:

$$\rho(T, H) = \frac{2\rho_c U}{T} e^{\frac{-U}{T}} \tag{5.6}$$

where U is the thermally activated energy. Assuming that $\rho_0 = 2\rho_0 U/T$ is a constant which is independent on temperature, and $U = U_0(1 - T/T_c)$, the above equation can be written in the Arrhenius relation:

$$\ln\rho(T, H) = \ln\rho_0 + \frac{U_0}{T_c} - \frac{U_0}{T} \tag{5.7}$$

As a result, the pinning energy at 0 K can be derived from the slope of the lnρ-T^{-1} curves. Figure 5.13 show the Arrhenius plots of resistivity as lnρ-T^{-1} for flat rolled and hot pressed Sr-122 tapes measured in perpendicular fields up to 9 T. The Arrhenius relation holds well in a wide temperature range, indicating that the TAFF model fits well to the flux motion of the Sr-122 tapes. The flux pinning energy can be derived from the slope $U_0(H) = -\mathrm{d}\ln\rho/\mathrm{d}\,T^{-1}$.

The thermally activation energy U_0 range from 4955 K at 1 T to 2797 K at 9 T for the flat rolled tapes and from 6024 K at 1 T to 3551 K at 9 T for the hot pressed tapes, as shown in Fig. 5.14. It can be seen that $U_0(H)$ follows a power law [i.e., $U_0(H) \sim H^{-\alpha}$] as a function of magnetic field for both flat rolled and hot pressed Sr-122 tapes.

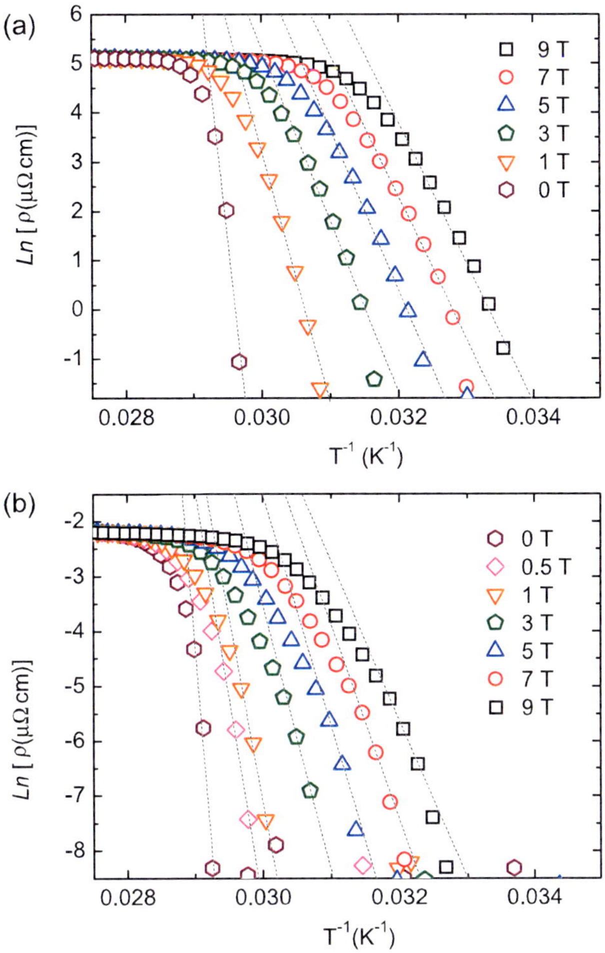

Fig. 5.13 Arrhenius plots of resistivity at various magnetic fields perpendicular to the tape plane for **a** flat rolled tapes and **b** hot pressed Sr-122 tapes

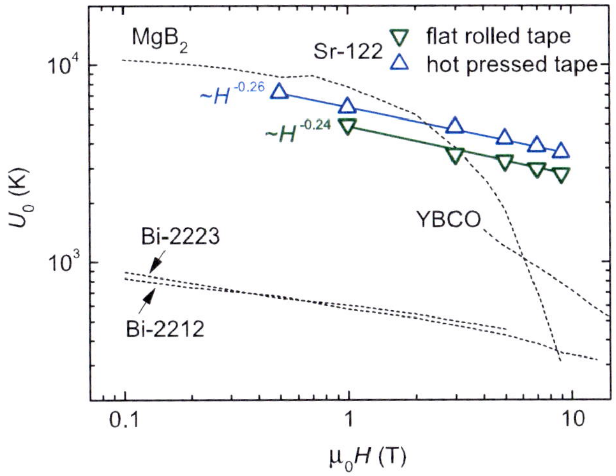

Fig. 5.14 Field dependence of flux pinning energy U_0 for flat rolled and hot pressed Sr-122 tapes. The applied field is perpendicular to the tape plane. The solid lines are the fitting results using $U(\mu_0 H) \sim AH^{-\alpha}$. Field dependence of flux pinning energy U_0 for MgB_2, YBCO, Bi-2223 and Bi-2212 materials is shown using the dashed lines for comparison

On the other hand, the activation energy exhibits constant power-law dependences on magnetic field, which can be scaled with $\alpha = 0.24$ and 0.26, respectively. These scale values are comparable to that of Ba-122 single crystal, which is almost field independent [39]. For comparison, flux pinning energy U_0 for MgB_2, YBCO, Bi-2223 and Bi-2212 materials is also inclided in Fig. 5.14. The pinning energy of the Sr-122 samples is comparable to that reported for MgB_2 (8700 K) and significantly larger than that reported for Bi-2212 (700 K) and Bi-2223 (650 K) measured under the same field [38, 40, 41]. Furthermore, the Sr-122 tapes exhibit a higher pinning energy than MgB_2 and YBCO in the field of $H > 3$ T [42], and larger than BSCCO in the whole magnetic field. In the high-field refion ($H > 3$ T), the field dependence of pinning energy for Sr-122 tapes is much weaker than that for MgB_2 and YBCO supercoductors. In a word, the Sr-122 tapes exhibt ultrahigh upper critical fields and very high pinning energy with weak field dependence, which is the intrinsic origin of for the high J_c performance in high magnetic fields.

5.3 Conclusion

The investigate for the mechanism of J_c enhancement for 122-type IBS tapes, including superconducting properties and microstructure analysis, reveal that the high J_c achieved the Sr-122 tapes can be ascribed to the combination of good grain connectivity, high degree of texture, and strong pinning strength. With MOI technique, magnetic flux distribution of the superconducting cores in Meissner state and remnant state was characterized. The temperature dependence of the J_c values measured by using the magnetic methods and transport methods are comparatively studied. The results revealed the remarkably improved grain connectivity in flat rolled Sr-122 tapes, as well as inhomogeneous current carrying ability along the length of the tapes. With the HRTEM, lots of textured grains with clean and low-angle grain boundaries were observed in hot pressed Sr-122 tapes. Therefore, grain alignment also plays an important role in the contribution to the percolation behavior of the current path in the Sr-122 tapes. Moreover, strong flux pinning ability is another basis for superior J_c-B performance. It is found that the flux motion in Sr-122 tapes can be well explained by the thermal activated flux flow model, with which high pinning energy with weak field dependence is derived for both flat rolled and hot pressed samples, rendering 122-type IBS to be interesting for high-field magnet applications.

References

1. Ma Y (2012) Progress in wire fabrication of iron-based superconductors. Supercond Sci Technol 25:113001

2. Hosono H, Yamamoto A, Hiramatsu H, Ma Y (2018) Recent advances in iron-based superconductors toward applications. Mater Today 21:278
3. Wang L, Qi Y, Wang D, Zhang X, Gao Z, Zhang Z, Ma Y, Awaji S, Nishijima G, Watanabe K (2010) Large transport critical currents of powder-in-tube $Sr_{0.6}K_{0.4}Fe_2As_2$/Ag superconducting wires and tapes. Phys C 470:183
4. Wang L, Qi Y, Zhang Z, Wang D, Zhang X, Gao Z, Yao C, Ma Y (2010) Influence of Pb addition on the superconducting properties of polycrystalline $Sr_{0.6}K_{0.4}Fe_2As_2$. Supercond Sci Technol 23:054010
5. Yao C, Wang C, Zhang X, Wang L, Gao Z, Wang D, Wang C, Qi Y, Ma Y, Awaji S, Watanabe K (2012) Improved transport critical current in Ag and Pb co-doped $Ba_xK_{1-x}Fe_2As_2$ superconducting tapes. Supercond Sci Technol 25:035020
6. Cai XY, Polyanskii A, Li Q, Riley GN Jr, Larbalestier DC (1998) Current-limiting mechanisms in individual filaments extracted from superconducting tapes. Nature 392:906
7. Koblischka MR, Johansen TH, Bratsberg H, Vase P (1998) Study of flux entry and exit into Bi-2223 multifilamentary tapes. Supercond Sci Technol 11:479
8. Bobyl AV, Shantsev DV, Johansen TH, Baziljevich M, Galperin YM, Gaevski ME (2000) Temperature dependence of filament-coupling in Bi-2223 tapes: magneto-optical study. Supercond Sci Technol 13:183
9. Feldmann DM, Reeves JL, Polyanskii AA et al (2000) Influence of nickel substrate grain structure on $YBa_2Cu_3O_7$ supercurrent connectivity in deformation-textured coated conductors. Appl Phys Lett 77:2906
10. Polyanskii AA, Feldmann DM, Patnaik S et al (2001) Examination of current limiting mechanisms in monocore $Bi_2Sr_2Ca_2Cu_3O_x$ tape with high critical current density. IEEE Trans Appl Supercond 11:3269
11. Tamegai T, Nakajima Y, Tsuchiya Y, Iyo A, Miyazawa K, Shirage PM, Kito H, Eisaki H (2009) Magneto-optical imaging of iron-oxypnictide $SmFeAsO_{1-x}F_x$ and $SmFeAsO_{1-y}$. Physica C 469:915
12. Prozorov R, Tillman M, Mun E, Canfield P (2009) Intrinsic magnetic properties of the superconductor $NdFeAsO_{0.9}F_{0.1}$ from local and global measurements. New J Phys 11:035004
13. Nakajima Y, Tsuchiya Y, Taen T, Tamegai T, Okayasu S, Sasase M (2009) Enhancement of critical current density in co-doped $BaFe_2As_2$ with columnar defects introduced by heavy-ion irradiation. Phys Rev B 80:012510
14. Taen T, Tsuchiya Y, Nakajima Y, Tamegai T (2009) Superconductivity at $T_c \sim 14$ K in single crystalline $FeTe_{0.61}Se_{0.39}$. Phys Rev B 80:092502
15. Sun Y, Taen T, Tsuchiya Y, Shi ZX, Tamegai T (2013) Effects of annealing, acid and alcoholic beverages on $Fe_{1+y}Te_{0.6}Se_{0.4}$. Supercond Sci Technol 26:015015
16. Ding Q, Prombood T, Tsuchiya Y, Nakajima Y, Tamegai T (2012) Superconducting properties and magneto-optical imaging of $Ba_{0.6}K_{0.4}Fe_2As_2$ PIT wires with Ag addition. Supercond Sci Technol 25:035019
17. Weiss JD, Tarantini C, Jiang J, Kametani F, Polyanskii AA, Larbalestier DC, Hellstrom EE (2012) High intergrain critical current density in fine-grain $(Ba_{0.6}K_{0.4})Fe_2As_2$ wires and bulks. Nat Mater 11:682
18. Wang C, Yao C, Lin H, Zhang X, Zhang Q, Wang D, Ma Y, Awaji S, Watanabe K, Tsuchiya Y, Sun Y, Tamegai T (2013) Large transport J_c in Sn-added $SmFeAsO_{1-x}F_x$ tapes prepared by an ex situ PIT method. Supercond Sci Technol 26:075017
19. Tamegai T, Ding QP, Inoue H, Taen T, Tsuchiya Y, Mohan S, Sun Y, Prombood T, Nakajima Y (2013) Magneto-optical characterization of iron-based superconducting wires and tapes. IEEE Trans Appl Supercond 23:7300304
20. Yao C, Wang C, Zhang X, Wang D, Lin H, Zhang Q, Ma Y, Tsuchiya Y, Sun Y, Tamegai T (2014) A comparative study of $Sr_{1-x}K_xFe_2As_2$ and $SmFeAsO_{1-x}F_x$ superconducting tapes by magneto-optical imaging. Supercond Sci Technol 27:044019
21. Sun Y, Taen T, Tsuchiya Y, Ding QP, Pyon S, Shi ZX, Tamegai T (2013) Large, homogeneous, and isotropic critical current density in oxygen-annealed $Fe_{1+y}Te_{0.6}Se_{0.4}$ single crystal. Appl Phys Express 6:043101

22. Frankel DJ (1979) Critical state model for the determination of critical currents in disk-shaped superconductors. J Appl Phys 50:5402
23. Daumling M, Larbalestier DC (1989) Critical state in disk-shaped superconductors. Phys Rev B 40:9350
24. Brandt EH, Indenbom MV, Forkl A (1993) Type-II superconducting strip in perpendicular magnetic field. Europhys Lett 22:735
25. Bean CP (1964) Magnetization of high-field superconductors. Rev Mod Phys 36:31
26. Wang L, Ma Y, Wang Q, Li K, Zhang X, Qi Y, Gao Z, Zhang X, Wang D, Yao C, Wang C (2011) Direct observation of nanometer-scale amorphous layers and oxide crystallites at grain boundaries in polycrystalline $Sr_{1-x}K_xFe_2As_2$ superconductors. Appl Phys Lett 98:222504
27. Zhang X, Wang Q, Li K, Cai Y, Jiang F, Wang Z, Li J, Yao C, Lin H et al (2015) Investigation of J_c-suppressing factors in flat-rolled $Sr_{0.6}K_{0.4}Fe_2As_2$/Fe Tapes. IEEE Trans Appl Supercond 25:7300105
28. Katase T, Ishimaru Y, Tsukamoto A, Hiramatsu H, Kamiya T, Tanabe K, Hosono H (2011) Advantageous grain boundaries in iron-pnictide superconductors. Nat Commun 2:409
29. Gurevich A, Pashitskii EA (1998) Current transport through low-angle grain boundaries in high-temperature superconductors. Phys Rev B 57:13878
30. Zhang X, Yao C, Lin H, Cai Y, Chen Z, Li J, Dong C, Zhang Q, Wang D, Ma Y, Oguro H, Awaji S, Watanabe K, Realization of practical level current densities in $Sr_{0.6}K_{0.4}Fe_2As_2$ tape conductors for high-field applications. Appl Phys Lett 104:202601
31. Hensel B, Grasso G, Flukiger R (1995) Limits to the critical transport current in superconducting $(Bi, Pb)_2Sr_2Ca_2Cu_3O_{10}$ silver-sheathed tapes: the railway-switch model. Phys Rev B 51:15456
32. Yan Y, Evetts JE, Soylu B, Stobbs WM (1996) The origins of high values of the critical current density in the $Bi_2Sr_2Ca_{n-1}Cu_nO_x$ system: high-T_c superconducting pathways at low angle tilt boundaries. Physica C 261:56
33. Feng Y, Larbalestier DC (1994) Texture relationships and interface structure in Ag-sheathed Bi(Pb)–Sr–Ca–Cu–O superconducting tapes. Interface Sci 1:401–410
34. Eibl O (1995) The high-T_c compound $(Bi, Pb)_2Sr_2Ca_2Cu_3O_{10+\delta}$: features of the structure and microstructure relevant for devices in magnet and energy technology. Supercond Sci Technol 8:833–861
35. Hiramatsu H, Katase T, Ishimaru Y, Tsukamoto A, Kaniya T, Tanabe K, Hosono H (2012) Microstructure and transport properties of [0 0 1]-tilt bicrystal grain boundaries in iron pnictide superconductor cobalt-doped $BaFe_2As_2$. Mater Sci Eng B 177:515
36. Godeke A, Jewell MC, Fischer CM, Squitieri AA, Lee PJ, Larbalestier DC (2005) The upper critical field of filamentary Nb_3Sn conductors. J Appl Phys 97:093909
37. Tarantini C, Gurevich A, Jaroszynski J, Balakirev F, Bellingeri E, Pallecchi I, Ferdeghini C, Shen B, Wen HH, Larbalestier DC (2011) Significant enhancement of upper critical fields by doping and strain in iron-based superconductors. Phys Rev B 84:184522
38. Palstra T, Batlogg B, Schneemeyer LF, Waszczak JV (1988) Thermally activated dissipation in $Bi_{2.2}Sr_2Ca_{0.8}Cu_2O_{8+\delta}$. Phys Rev Lett 61:1662
39. Wang XL, Ghorbani SR, Lee S, Dou SX, Lin CT, Johansen TH, Mueller KH, Cheng ZX, Peleckis G, Shabazi M, Qviller AJ, Yurchenko VV, Sun GL, Sun DL (2010) Very strong intrinsic supercurrent carrying ability and vortex avalanches in $(Ba, K)Fe_2As_2$ superconducting single crystals. Phys Rev B 82:024525
40. Sidorenko A, Zdravkov V, Ryazanov V, Horn S, Klimm S, Tidecks R, Wixforth A, Koch TH, Schimmel TH (2005) Thermally assisted flux flow in MgB_2: strong magnetic field dependence of the activation energy. Philos Mag 85:1783
41. Wang XL, Li AH, Yu S, Ooi S, Hirata K, Lin CT, Collings EW, Sumption MD, Bhatia M, Ding SY, Dou SX (2005) Thermally assisted flux flow and individual vortex pinning in $Bi_2Sr_2Ca_2Cu_3O_{10}$ single crystals grown by the traveling solvent floating zone technique. J Appl Phys 97:10B114
42. López D, Krusin-Elbaum L, Safar H, Righi E, de la Cruz F, Grigera S, Field C, Kwok WK, Paulius L, Crabtree GW (1998) Pinned vorted liquid above the critical point of the first-order melting transition: a consewuence of pointlike disorder. Phys Rev Lett 80:1070

Chapter 6
Development of Multifilamentary 122-Type IBS Tapes

For practical applications, a conductor is more than just the superconductor [1]. Except for large current carrying capacity, they should be fabricated with superconducting multifilaments embedded in metal matrix for the protection against AC loss, flux jumps and thermal quenching. Moreover, a conductor is required to have sufficient mechanical strength to withstand the electromagnetic and thermal stress during operation. In addition, the product cost of is also an important issue which should be taken into consideration for large-scale commercial applications. Therefore, developing high-strength low-cost multifilament wires and tapes is essential to promote the application of IBS materials.

For the multifilamentary fabrication, it can be learned from the history of Bi-based cuprate (BSCCO) superconducting wires and tapes, since they are also prepared based on powder-in-tube technology. Starting powders, such as oxides and carbonates, are mixed and calcined to obtain precursors that are packed into a metal sheath, which is then mechanically deformed into wires and tapes. Ag or Ag alloy is required as sheath material instead of Cu and other metals for BSCCO wires and tapes, because they are chemically compatible (i.e., Ag hardly reacts with the precursor) and transparent to oxygen (i.e., Ag can transmit the oxygen released from or absorbed into oxide superconductors during heat treatment). Multifilamentary wires can be fabricated by further restacking monofilamentary wires in an Ag sheath and then repeating the cold deformation process.

At present, though high transport J_c has been obtained in Ag-sheathed IBS tapes, there are two drawbacks of Ag sheaths that will hold back the large-scale applications. One is the high price due to the use of noble silver. As for the case of Bi-2223 tapes, the price-performance ($/kAm, at 4 K and 6 T) of Bi-2223 tapes is about 60–80 $/kAm, which is still much higher than the ideal 25 $/kAm for large-scale applications [2]. The other drawback of Ag sheath is its poor mechanical properties, which can hardly provide sufficient protection for the superconducting filaments inside wires and tapes. The weak mechanical strength of wires and tapes usually require a wind-and-react

C. Yao, *Fabrication and Properties of High-Performance 122-Type Iron-Based Superconducting Wires and Tapes*, Springer Theses,
https://doi.org/10.1007/978-981-10-5184-5_6

method for magnet constructions, other than the more convenient and flexible react-and-wind method. With laminated mechanical reinforcement technique by Sumitomo for the weak Ag sheathed Bi-2223 tapes, the mechanical strength of the tapes can be enhanced to meet the requirement of high-field applications [3], but on the other hand this process will lower the filling factor of the whole tapes and bring extra cost.

For IBS wires and tapes, since the oxygen permeability of sheath material is not needed during sintering process, the material choice for the sheath is not limited to silver or some silver rich alloys, using other cheap and stiff metal material as the outer sheath combined with Ag inner sheath can be a practical proposal, which can enhance mechanical properties, reduce the ratio of Ag, and provide chemical stability at the same time for IBS conductors. Therefore, high-performance multifilamentary IBS wire and tapes with composite sheaths are very promising for future applications. In this chapter, based on PIT process, high-performance 7- and 19-filament Ag sheathed 122-IBS multifilamentary wires and tapes were fabricated by drawing and flat rolling process. Moreover, Fe/Ag composite sheathed multifilamentary wires and tapes with filaments up to 114 were developed.

6.1 7-Filament Sr-122 Wires and Tapes

The 7-filament wires and tapes can be made by one-time restacking 7 single-filament wires into a metal tube and then cold-worked into multifilamentary wires and tapes. The process for synthesizing Sr-122 precursor is the same as used for single-filament wires and tapes. The carefully ground Sr-122 powder was packed into Ag tubes (OD: 8 mm and ID: 5 mm), which were drawn into wires of 1.58 mm diameter with a reduction rate of about 5% and cut into short wires. As shown in Fig. 6.1,

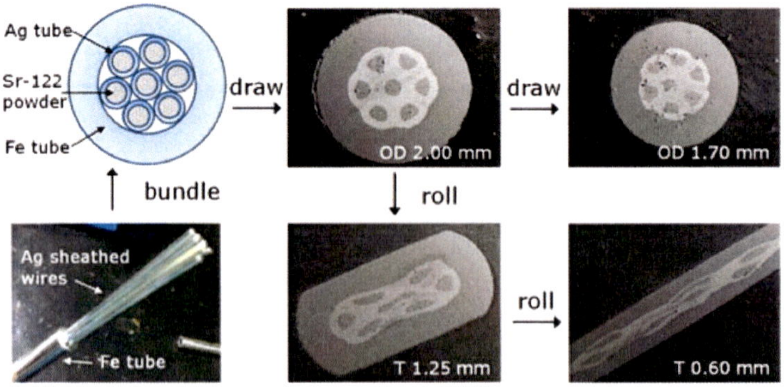

Fig. 6.1 Fabrication process of multifilamentary Sr-122 wires and tapes with various diameters and thicknesses. SEM images of the cross section of wires (diameter = 2.00 and 1.70 mm) and tapes (thickness = 1.25 and 0.60 mm) are shown. Reprint from Yao et al. [4]

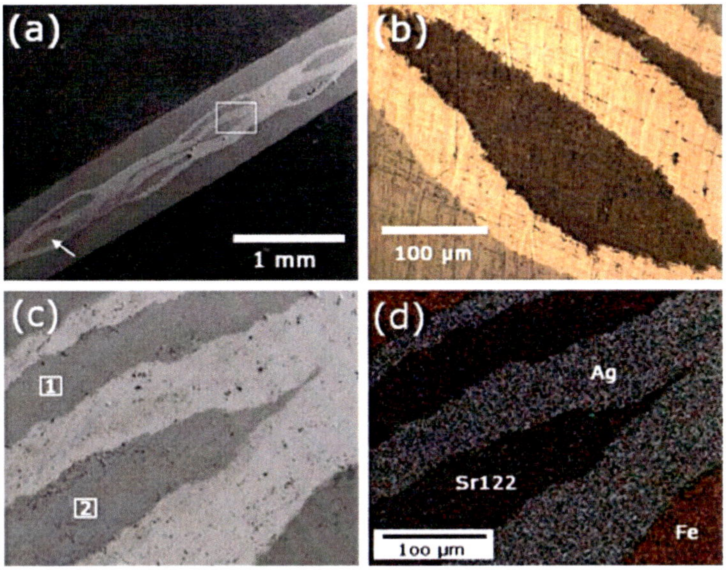

Fig. 6.2 **a** SEM image presenting the transverse cross section of the 0.60 mm thick 7-filament Sr-122 tape; **b** Optical microscope image presenting one of the seven Sr-122 filaments marked by the arrow in this figure; **c** SEM image and **d** its EDX elemental mapping of the selected area marked by the white rectangular frame in this figure. Reprint from Yao et al. [4]

seven short Ag sheathed wires were then bundled into a Fe tube (OD: 8 mm and ID: 5 mm), and drawn into wires of about 2.00, 1.90, 1.80 and 1.70 mm diameters with a reduction rate of about 5%. Some multifilamentary wires of 2.00 mm diameter were successively rolled into tapes of about 1.50, 1.25, 1.00, 0.80 and 0.60 mm thicknesses. The obtained wires and tapes were cut into short samples and then sealed into quartz tubes, and heat treated at 850 ~ 950 °C for about 0.5 h.

The SEM images in Fig. 6.1a show the cross section of well-developed 7-filament Sr-122 wires and tapes. Even when the as-drawn wires were rolled into tapes of 0.6 mm thickness, the Ag inner sheaths were found to still well separate the Sr-122 superconducting filaments. As shown in Fig. 6.1b, an optical image taken by a microscope at the area marked by the arrow in Fig. 6.1a shows one of the seven filaments in the 0.60 mm thick tape. It can be seen that the Sr-122 filament looks very dense, and the interface between the Sr-122 phase and Ag inner sheath is very clear, indicating the absence of nonsuperconducting reaction layer after heat treatment. The area marked by the white rectangular frame in Fig. 6.1a was studied by means of EDX. Comparing the original image taken by SEM in Fig. 6.1c with the corresponding EDX element mapping in Fig. 6.1d, it is obvious that the areas of Sr-122, Ag sheath and Fe sheath in these two images are well accordant and tightly connected. In order to determine the actual composition of the Sr-122 filaments, two of the filaments were also analyzed with EDX, and the selected areas were marked with white rectangular frames in Fig. 6.1a. The result of Area 1 and 2 is $Sr_{0.59}K_{0.41}Fe_2As_{1.99}$

and $Sr_{0.59}K_{0.37}Fe_2As_{2.04}$ respectively. Though the amount of K is slightly inhomogeneous, this result is very close to the nominal composition. These results suggest that the Ag/Fe sheath is suitable for fabricating Sr-122 multifilamentary wires and tapes.

The areas of the whole cross section ($S_{Fe+Ag+Sr122}$) and Sr-122 filaments (S_{Sr122}) for the wires with different diameters and tapes with different thicknesses are listed in Table 6.1. As calculated in Table 6.1, the ratios of S_{Sr122} to $S_{Fe+Ag+Sr122}$ are not obviously changed during the drawing and rolling process. However, there is somewhat inhomogeneity between different filaments during the rolling deformation process. The change of the area of each Sr-122 filament is summarized in Fig. 6.3. It can be seen that generally the areas of all filaments decrease with the reduction of tape

Table 6.1 Cross-sectional areas for the whole samples and superconducting filaments of the 7-filament Sr-122 wires and tapes

Samples	Diameter/thickness (mm)	$S_{Fe+Ag+Sr122}$ (mm^2)	S_{Sr122} (mm^2)	$S_{Sr122}/S_{Fe+Ag+Sr122}$ (%)
Wire	2.00	3.299	0.290	8.79
	1.90	2.716	0.243	8.95
	1.80	2.487	0.208	8.36
	1.70	2.216	0.187	8.44
Tape	1.50	3.153	0.250	7.93
	1.25	3.149	0.249	7.91
	1.00	2.991	0.243	8.12
	0.80	2.863	0.227	7.93
	0.60	2.587	0.209	8.08

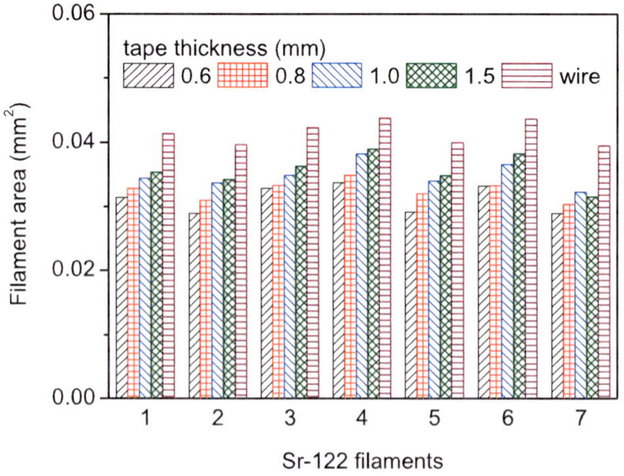

Fig. 6.3 Evolution of the transverse cross-sectional areas for the seven Sr-122 filaments during the flat rolling deformation process from wires into tapes of 1.5, 1.0, 0.8 and 0.6 mm in thickness

Table 6.2 Transport properties of the 7-filament Sr-122 wires and tapes

Samples	Diameter/thickness (mm)	$T_{c,\text{ onset}}$ (K)	$T_{c,\text{ zero}}$ (K)	ρ_{300} (μ Ω cm)	Transport J_c (kA/cm^2) (at 4.2 K, 0 T)
Wire	2.00	33.5	30.2	14.5	4.4
	1.90	33.7	30.3	13.9	4.1
	1.80	33.8	30.3	12.8	3.6
	1.70	33.6	31.4	10.7	4.4
Tape	1.50	33.1	28.0	13.1	2.2
	1.25	33.4	28.0	12.4	2.7
	1.00	33.3	29.8	11.7	2.7
	0.80	34.3	29.9	11.4	8.5
	0.60	34.5	30.4	10.1	21.1

thickness, but the decrease rates are not so accordant with each other, indicating the deformation is not uniform for the filaments at different locations on the cross section.

The transport properties including onset and zero superconducting transition temperature ($T_{c,\text{ onset}}$) and ($T_{c,\text{ zero}}$), resistivity at 300 K (ρ_{300}), and transport J_c of each sample are listed in Table 6.2. Here the resistivity ρ can be define by $S_{\text{Fe+Ag+Sr122}}/\rho = S_{\text{Fe}}/\rho_{\text{Fe}} + S_{\text{Ag}}/\rho_{\text{Ag}} + S_{\text{Sr122}}/\rho_{\text{Sr122}}$. Since the resistivity of Sr-122 single crystal is much larger than that of Ag and Fe at 300 K [5], and since the S_{Sr122} is just about 8 ~ 9% of $S_{\text{Fe+Ag+Sr122}}$, the value of ρ is mainly determined by the Ag and Fe sheath at this temperature. As listed in Table 6.2, the resistivity at 300 K for both wires and tapes decreases with the decrease of areas $S_{\text{Fe+Ag+Sr122}}$ and $S_{\text{Sr-122}}$. Therefore, the contact resistance between the Ag and Fe sheath should be also taken into consideration. With the reduction of wire diameters and tape thicknesses and the increase of deformation force, the contact between the Ag and Fe sheath tightens, thus reducing the electrical resistance and improving the transport properties.

The temperature dependence of the resistivity for both wire and tape samples are plotted in Fig. 6.4. For the wires with different diameters, the $T_{c,\text{ onset}}$ is almost the same during the deformation. On the other hand, the $T_{c,\text{ onset}}$ first slightly decrease after rolling the tapes with a rolling thickness of 1.50 mm, then seems to recover after further rolling the tapes to 1.25 and 1.00 mm in thickness, and finally increases for about 1 K in 0.80 and 0.60 mm thick tapes. It is known that the rolling can enhance the density of the superconducting fialments thus improving the grian connectivity, which is very critical to the tansport propeties. Moreover, the increase of density can improve the reaction and formation of Sr-122 phase, thus achieving a higher $T_{c,\text{ onset}}$. In this work, the homongenous fialments after drawing is somewhat damaged during transition from round wire to rectangular tape at the beginning, but then recovers and becomes even more dense by subsequent rolling. This can explain the change of $T_{c,\text{ onset}}$ in the rolling process that metioned above. Compared to the wires, the transition width of the tapes is more broader, indicating that the density

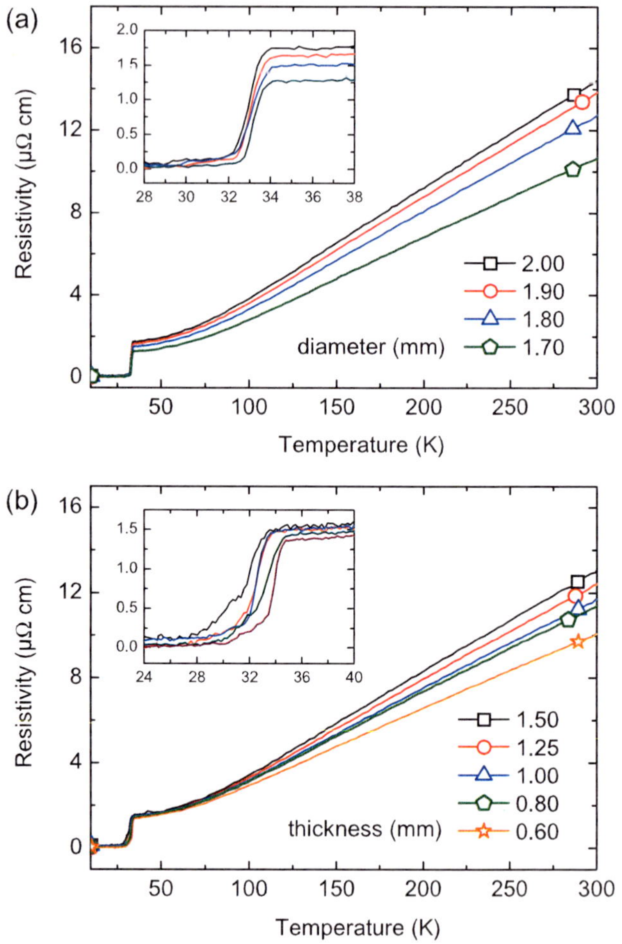

Fig. 6.4 Temperature dependence of resistivity for the **a** wires with various diameters and **b** tapes with various thicknesses. The insets show the enlarged part near the superconducting transition. Reprinted from Yao et al. [4]

becomes inhomogenous in Sr-122 filaments and is different among fialments at the different locations of the tapes, which is in accordance with the results in Fig. 6.3. The inhomogenity of superconducting fialments was also observed in the Bi-2223 multifilamentary tapes after cold-work denformation [6].

The effects of different deformation routes on the transport J_c prtformance can be discussed based on the data given in Table 6.2. Both starting with wires of 2.00 mm diameter, drawing has no significant effect on the until the wires are drawn to 1.70 mm in diameter, on the other hand, rolling first reduces the transport J_c for about 50% when deforming the wires into tapes, indicating the grain connectivity is weakened due to the deformation induced rearrangement of mass density in the Sr-122 filaments. When the tapes were rolled thinner than 1.0 mm, the transport J_c show a significant enhancement, and finally a J_c about 5 times as high as the staring wire is achieved in the 0.60 mm thick tapes. This can be ascribed to the fully improved grain connectivity and grain texture. The evolution of the transport J_c at 4.2 K in self-field with various wire diameters and tape thicknesses is presented in Fig. 6.5.

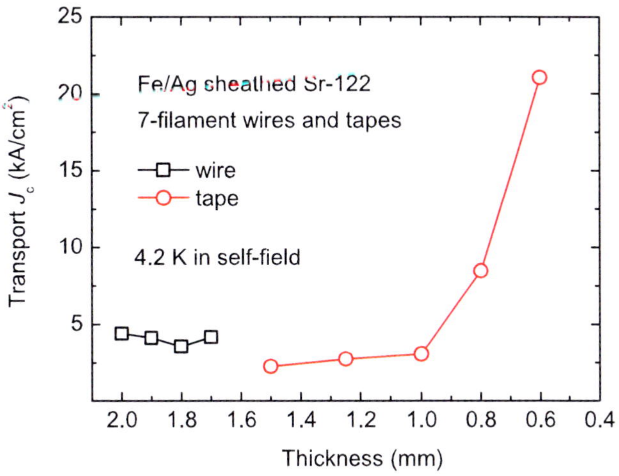

Fig. 6.5 Evolution of the transport J_c at 4.2 K in self-field for the Fe/Ag sheathed 7-filament Sr-122 wires and tapes during the deformation process

It can be noted that the change of transport J_c at 4.2 K in self field is generally in agreement with that of $T_{c,\,\mathrm{onset}}$, which can also be ascribe to the change of mass density caused by rolling. Therefore, it can be concluded that compared to drawing, rolling is more effective in enhancing the Sr-122 density thus achieving higher J_c in 7-filament tapes. In the fabrication process of mono- and multifilamentary Bi-2223 tapes, Bi-2212 tapes and MgB_2 tapes [6–11] the rolling induced increase of density plays an important role in the improvement of transport current, similarly to the Sr-122 tapes in this work.

Figure 6.6 presents the field dependent transport J_c of tapes rolled to 1.00, 0.80 and 0.60 mm in thickness. It is found that the transport J_c significantly increases in the whole field region when reducing the rolling thickness. The J_c values are the highest for the 0.60 mm thick tapes and achieve 2.1×10^4 A/cm^2 at 4.2 K in self field and 3.3×10^3 A/cm^2 at 10 T. It decreases with the increasing field to about 0.5 T, showing

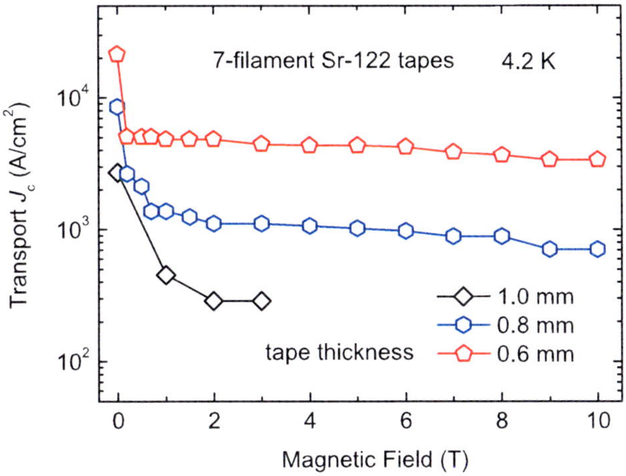

Fig. 6.6 Field dependence of the transport J_c of 7-filament Sr-122 tapes (thickness = 1.0, 0.8 and 0.6 mm). The applied field is parallel to the tape plane

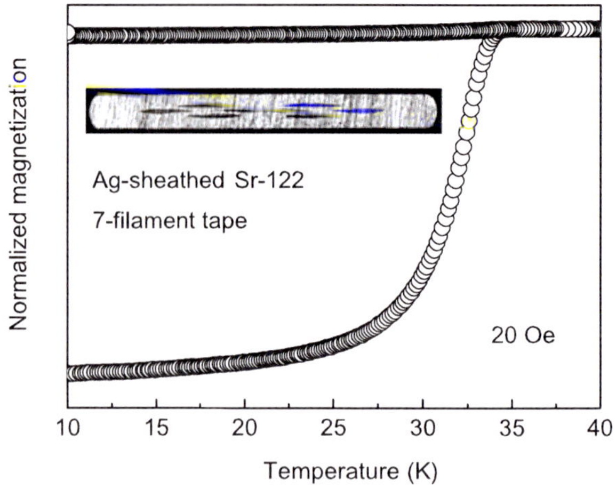

Fig. 6.7 Temperature dependence of normalized magnetization of the Ag-sheathed Sr-122 7-filament tape measured in zero-field (ZFC) and field cooling (FC) procedures

very weak field dependence up to 10 T. By optimizing the cold-work parameters to obtain denser microstructure, it is possible to further improve the transport J_c in the future. Moreover, considering the extremely weak field dependence of their transport J_c, the multifilamentary wires of 122-type iron-based superconductors will be promising candidates in high-field applications.

Since there is a certain degree of nonuniform deformation for the Fe/Ag sheathed multifilamentary tapes as described above, Ag-sheathed 7-filament tapes were also fabricated to evaluate the effect of Fe/Ag composite sheath on the J_c-performance of multifilamentary tapes. The fabricating parameters for Ag-sheathed tapes are similar to that used to prepare Fe/Ag sheathed multifilamentary tapes, and the major difference is the materials of the sheaths. In addition, except for annealing in sealed quartz tubes, some tape samples were annealed with hot pressing process. The transverse cross section of Ag-sheathed 7-filament tapes with common heat treatment is shown in Fig. 6.7. The Sr-122 filaments of Ag-sheathed sample show flatter and more regular shape, indicating more uniform cold-work deformation.

The normalized *M-T* curve with zero-field and field cooling of the tape sample is also presented in Fig. 6.7, showing a clear superconducting transition strating at 34.3 K. The field dependence of transport J_c at 4.2 K for flat rolled and hot pressed 7-filament Sr-122 tapes is shown in Fig. 6.8, respectively. The transport J_c at 10 T for the flat rolled tapes with common heat treatment is 2.1×10^4 A/cm^2, which is about 6 times higher than that for the Fe/Ag sheathed tapes. Processed with hot pressing, the Ag-sheathed 7-filament tapes sustain J_c as high as 6.1×10^4 A/cm^2 at 10 T, which is the best record in multifilamentary iron-based superconducting wires and tapes.

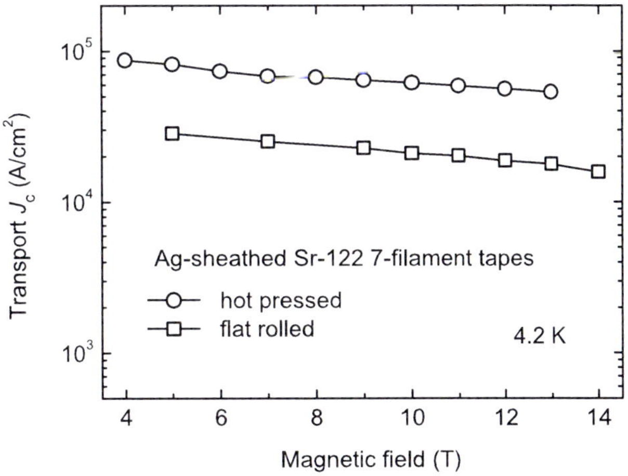

Fig. 6.8 Field dependence of transport J_c at 4.2 K of flat rolled and hot pressed 7-filament Sr-122 tapes. The applied field is parallel to the tape plane

6.2 19-Filament Sr-122 Wires and Tapes

With a similar fabrication process to the above mentioned 7-filament Sr-122 wires and tapes, 19-filament Sr-122 superconducting wires and tapes with silver as matrix and iron as outer reinforcing sheath were produced, as shown in Fig. 6.9. In order to check the crystallinity of Sr-122 phase, the superconducting transition for 19-filament Sr-122 wires and tapes was examined by *R-T* measurements from 10 to 300 K, as presented in Fig. 6.10. The onset and zero superconducting transition temperature ($T_{c,\text{ onset}}$ and $T_{c,\text{ zero}}$) is 33.9 and 32.6 K for wire of 2.0 mm in diameter and 34.9 and 33.4 K for tape of 0.6 mm in thickness. All the wire and tape samples show very steep superconducting transitions with transition widths smaller than 1.5 K. Compared with the wire sample, the 0.6 mm thick tape sample shows an increase of ~1.0 K

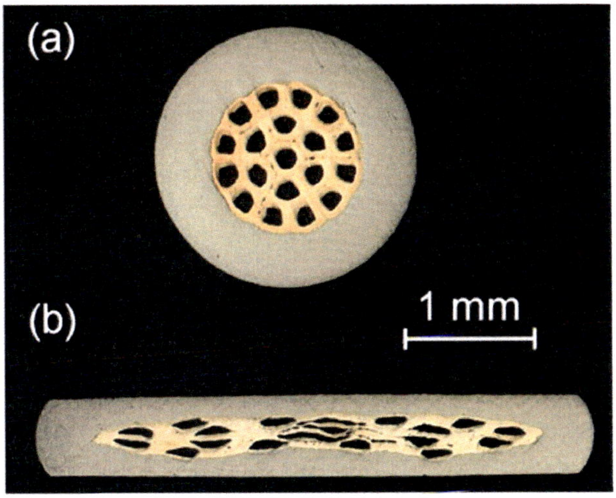

Fig. 6.9 Optical images of the transverse cross section for 19-filament Sr-122/Ag/Fe wire of 2.0 mm in diameter and tape of 0.6 mm in thickness

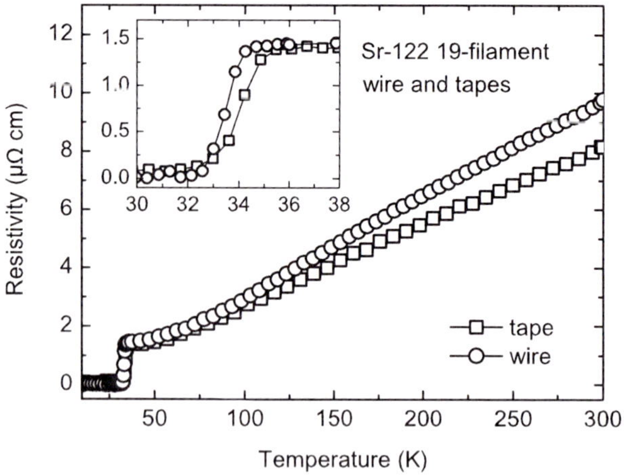

Fig. 6.10 Temperature dependence of the resistivity for 19-filament Sr-122/Ag/Fe wire of 2.0 mm in diameter and tape of 0.6 mm in thickness. The insets show the enlarged part near the superconducting transition

for both $T_{c,\ \mathrm{onset}}$ and $T_{c,\ \mathrm{zero}}$, indicating the deformation induced enhancement for the density of Sr-122 phase, since a highly dense superconducting phase is beneficial to its reaction and formation during the heat treatment, thus results in a high T_c. Similar to the case of 7-filament samples, the 19-filament tape sample exhibits a lower resistivity in the normal state than that of the wire sample, which is the result of improved contact of silver and iron sheath materials.

The field dependence of the transport J_c for 19-filament Sr-122 tapes with various thicknesses is plotted in Fig. 6.11. The transport J_c gradually increases with the reduction of tape thickness, and finally the J_c value comes up to 8.4×10^3 A cm^{-2} (4.2 K, 10 T) in 0.6 mm thick tapes. The field dependence of the transport J_c for 19-filament Sr-122 tapes is similar to that of 7-filament tapes. There is a dramatic drop for the transport J_c when the applied field was incresed to 1 T, but when the applied field was further increased to higher level up to 14 T, the decrease of transport J_c becomes quite slow. The sudden drop of transport J_c in the low field region indicates

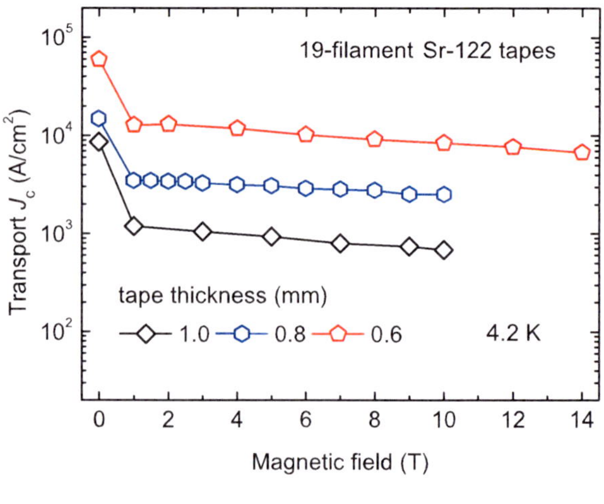

Fig. 6.11 Field dependence of the transport J_c of 19-filament Sr-122 tapes (thickness = 1.0, 0.8 and 0.6 mm). The applied field is parallel to the tape plane

that there are considerable weak-connected grains in the Sr-122 filaments, and the dope is less dramatic for the tapes with a lower thickness, which may be the result of increased Sr-122 mass density.

In order to study the change of mass density of the Sr-122 filaments in the wires and tapes made by different cold-work parameters, Vickers hardness was measured for the Sr-122 filaments. Being much more convenient than the dierect measurement of mass density for superconducting fialments, hardness can be used to monitor the evolution of local mass density during the manufacturing process [6]. The 19-filament Sr-122/Ag/Fe round wire shows relatively uniform Sr-122 densities with Vickers hardness of 38 ~ 69, respectively. Due to the defomation induced densification, the 19-filament tapes flat rolled to 0.6 mm in thickness show much higher hardness compared with round wires. The lowest value of hardness for 19-filament tape was measured in the filaments in the center area of the tapes, and the filaments near the silver sheath have higher hardness up to 133. This result is similar to the flat rolled 7-filament BiSrCaCuO/Ag tapes, in which the filament closer to the sheath show higher average hardness, though the microhardness profiles across each individual filament are different [12]. In addition, the hardness for silver matrix and iron sheath for all the samples was measured, and the results show no significant difference for all the samples. The hardness for silver matrix is 30 ~ 39, similar to the values in sintered Bi-2223/Ag tapes [13], and the hardness for iron sheath is 91 ~ 97, higher than silver matrix. Therefore, the mechanical strength for the whole conductors can be significantly improved by the outer iron sheaths.

Figure 6.12 presents the evolution of hardness for each Sr-122 filament in 19-filament wires and tapes and the average transport J_c (in self-field and 10 T) for the Sr-122 filaments when the conductors were rolled from round wires into flat tapes. It can be noted that the Sr-122 filaments exhibit different hardness between each other, which can be ascribed to the nonuniform cold-work deformation. On the other hand, in general their mass density and transport J_c both increase with

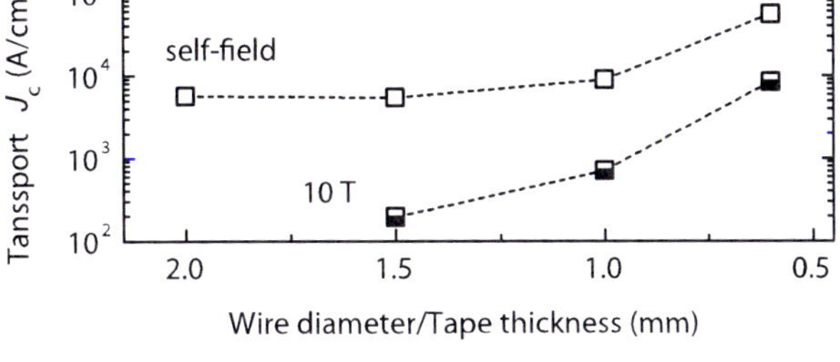

Fig. 6.12 Evolution of the hardness for Sr-122 filaments and their average transport J_c (at 4.2 K, in self-field and 10 T) with flat rolling for 19-filament Sr-122 wires and tapes

the reduction of tape thickness. This result is in accordance with the case of Ba-122/Ag and Ba-122/Ag/SS monofilamentary tapes, showing a positive correlation between hardness and transport J_c [14]. In addition, as reported in Sr-122 and Ba-122 monofilamentary tapes, whose grain orientation is usually improved along with the enhancement of density when the tapes are rolled or pressed thinner [15], the improvement of grain texture can also be a possible reason for the J_c increase of our flat rolled multifilamentaty tapes.

The flat rolled and hot pressed Ag-sheathed 19-filamentary tapes were also fabricated. The transverse cross section of Ag-sheathed 19-filament tapes with common heat treatment is shown in Fig. 6.13. Compared to the Ag-sheathed 7-filament tapes, the Sr-122 filaments are less regular in shape, indicating that it is more difficult to have a uniform deformation when the number of filaments is larger, but on the other hand they are more uniform that that for the Fe/Ag-sheathed 19-filament tapes shown in Fig. 6.9, especially for the filaments in the center part on the transverse cross section. Figure 6.13 also presents the normalized *M-T* curve with zero-field and field cooling of the tape sample, showing a clear superconducting transition strating at 32.3 K. The field dependence of transport J_c at 4.2 K for flat rolled and hot pressed 19-filament Sr-122 tapes is shown in Fig. 6.14, respectively. The transport J_c at 10 T for the flat rolled tapes with common heat treatment is 1.1×10^4 A/cm^2, which is also higher than that for the Fe/Ag sheathed tapes. However, the difference between the Fe/Ag- and Ag-sheathed tapes becomes much smaller in 19-filament tapes than in 7-filament tapes because of the improved precursor and fabricating process for the Fe/Ag sheathed tapes. After hot pressing, the transport J_c of 19-filament Ag-sheathed tapes at 10 T was significantly increased to 3.5×10^4 A/cm^2, which is about 3 times higher than that of the samples without hot pressing process. Such threefold increase is in accordance with the case for 7-filament Ag-sheathed tapes. Such high J_c values achieved hot pressed Ag sheathed multifilamentary tapes demonstrate that there is still much room for the J_c improvement for Fe/Ag sheathed multifilamentary tapes.

Fig. 6.13 Temperature dependence of normalized magnetization of the Ag-sheathed Sr-122 19-filament tape measured in zero-field (ZFC) and field cooling (FC) procedures

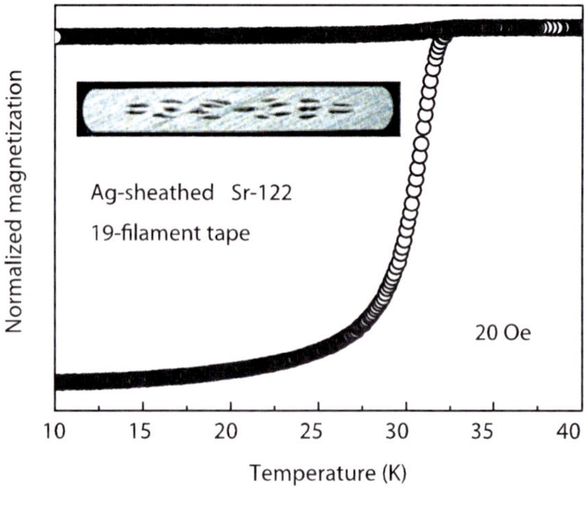

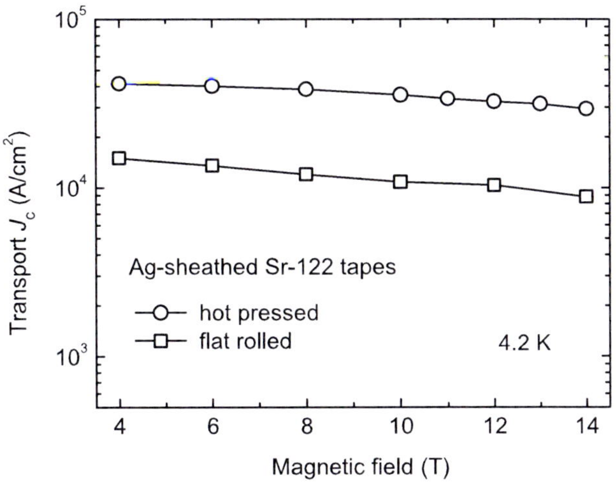

Fig. 6.14 Field dependence of transport J_c at 4.2 K of flat rolled and hot pressed 19-filament Sr-122 tapes. The applied field is parallel to the tape plane

6.3 114-Filament Sr-122 Wires and Tapes

For 114-filament Sr-122/Ag /Fe wires, 19 short 1.21 mm-diameter monofilamentary Sr-122/Ag wires were bundled into a silver tube (OD: 8 mm and ID: 6.4 mm), which was drawn into a 19-filament wire of 1.58 mm in diameter. Six short 19-filament Sr-122/Ag wires and a short silver wire (also with 1.58 mm in diameter) in the center were bundled into an iron tube (OD: 8 mm and ID: 5 mm), and finally deformed into 114-filament Sr-122/Ag/Fe wires of 2 mm in diameter and tapes with various thicknesses by drawing and flat rolling. The flow chart for the cold work process of the 114-filament wires and tapes is presented in Fig. 6.15. After cold deformation process, all the tape samples were submitted to a heat treatment at 900 °C for about 0.5 h. The transverse cross sections for the 114-filament Sr-122/Ag/Fe wires and tapes are presented in Fig. 6.16. As shown in Fig. 6.16a, filaments with size smaller than 50 μm were achieved in round wires with of 2.0 mm in diameter.

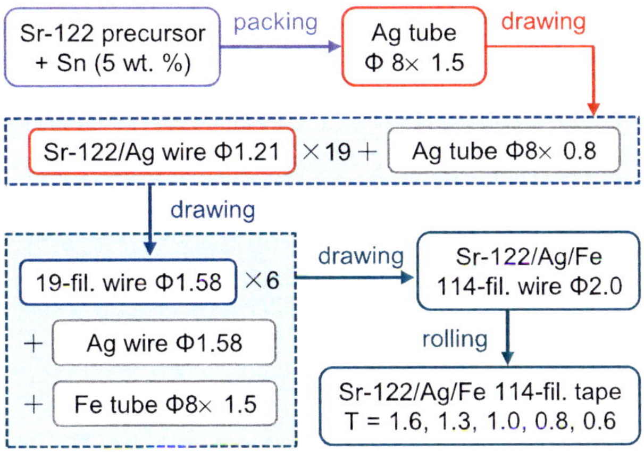

Fig. 6.15 Flow chart for the cold work process of the 114-filament Sr-122/Ag/Fe wires of 2.0 mm in diameter and tapes with various thicknesses

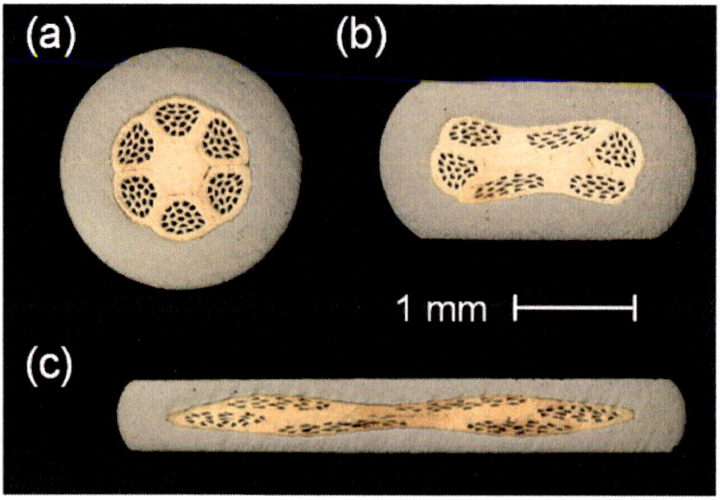

Fig. 6.16 Optical images of the transverse cross section for 114-filament Sr-122/Ag/Fe, **a** wires of 2.0 mm in diameter and tapes of **b** 1.0 mm and **c** 0.6 mm in thickness, showing the rearrangement of the Sr-122 filament during flat rolling. Reprint from Yao et al. [16]

The *R-T* curves from 10 to 300 K 114-filament Sr-122 wires and tapes were presented in Fig. 6.17. The onset and zero superconducting transition temperature ($T_{c,\ \mathrm{onset}}$ and $T_{c,\ \mathrm{zero}}$) is 33.9 and 32.6 K for wire of 2.0 mm in diameter and 35.1 and 33.9 K for tape of 0.6 mm in thickness. All the wire and tape samples show very steep superconducting transitions, and the transition widths are 1.2 ~ 1.3 K for them. Compared with the wire sample, the 0.6 mm thick tape sample shows an increase of 1.2–1.3 K for both $T_{c,\ \mathrm{onset}}$ and $T_{c,\ \mathrm{zero}}$. The increase of T_c for the 114-filament samples is slightly larger than that for 19-filament samples.

The field dependence of the transport J_c for 114-filament Sr-122/Ag/Fe wire of 2.0 mm in diameter and tapes with various thicknesses is plotted in Fig. 6.18. The transport J_c of round wires is about 800 A cm^{-2} (4.2 K, 10 T), and when they are flat rolled into tapes, the transport J_c gradually increseds with the reduction of the tape thickness. Finally the J_c value comes up to 6.3 × 10^3 A cm^{-2} (4.2 K, 10 T) in 0.6 mm thick tapes. Though the filaments are too small for hardness measurement for the 114-filament conductors, the mechanical deformation induced densification for the Sr-122 phase can be inferred since the transport J_c can be significantly enhanced by flat rolling as for the 7- and 19-filament conductors. It is noteworthy that these 114-filament samples shows almost the same quite weak field dependence up to 14 T, though they have different sizes and shapes, indicating very promising potential in high-field applications. This cold work independent J_c-H dependence is similar to the case of monofilamentary wires and tapes [13, 15], whose pinning behaviors are believed to be dominated by the intrinsic pinning of Sr-122 phase [17]. However, like the artificial defects introduced by irradiation in $Ba(Co,Fe)_2As_2$ thin films and SmFeAs(O,F) single crystals as pinning centers [18, 19], *c*-axis correlated pinning seems also to exist in the 122 tapes due to the rolling induced grain textures, according

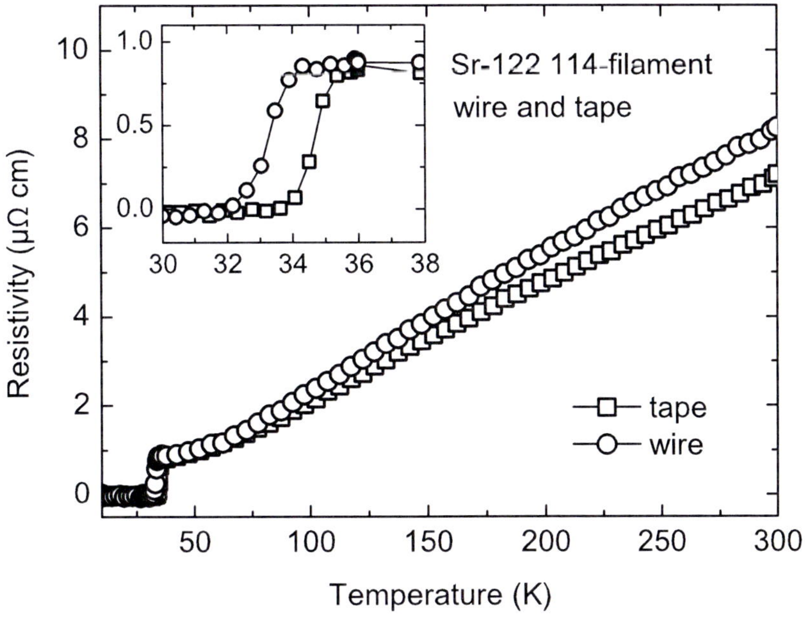

Fig. 6.17 Temperature dependence of the resistivity for 114-filament Sr-122/Ag/Fe wire of 2.0 mm in diameter and tape of 0.6 mm in thickness. The insets show the enlarged part near the superconducting transition

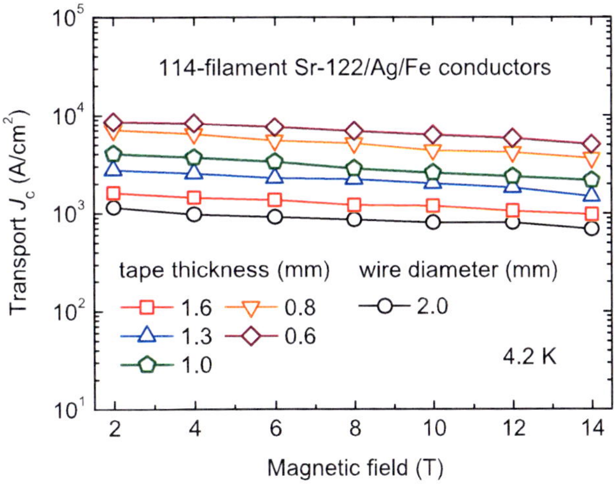

Fig. 6.18 Field dependence of the transport J_c for 114-filament Sr-122/Ag/Fe wire of 2.0 mm in diameter and tapes with various thicknesses. The applied field is parallel to the tape plane. Reprint from Yao et al. [16]

to the results of transport J_c as a function of the angle between magnetic fields directions and tape surfaces [20]. Therefore, detailed studies are still needed to understand the pinning mechanism of multifilamentary pnictide wires and tapes.

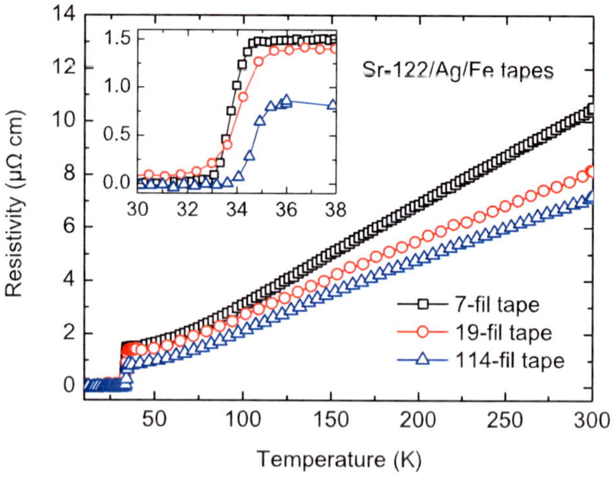

Fig. 6.19 Temperature dependence of the resistivity for 7-, 19- and 114-filament Sr-122/Ag/Fe tapes of 0.6 mm in thickness. The insets show the enlarged part near the superconducting transition

6.4 Comparative Study on 7, 19-and 114-Filament Sr-122 Tapes

With improved precursor and fabricating process, the transport J_c Fe/Ag sheathed Sr-122 7-filament tapes can be enhanced to 1.4×10^4 A cm^{-2} (4.2 K, 10 T), which is about 4 times higher than that of the 7-filament tapes in Sect. 6.1. The superconducting transition curves for 7-, 19- and 114-filament Sr-122/Ag/Fe tapes obtained by *R-T* measurements from 10 to 300 K are compared in Fig. 6.19. All the tape samples show $T_{c,\,\mathrm{onset}}$ above 34 K, and also very steep superconducting transitions with transition widths smaller than 1.5 K. Due to the increased ratio of sheath materials to Sr-122 phase, the resistivity in the normal state decreases with the increase of filament numbers. Based on the analysis of superconducting transition above, in this work the Sr-122 phase exhibits similar superconducting transitions in tape samples with different filament numbers. The field dependence of the transport J_c for 7-, 19- and 114-filament tapes is compared in Fig. 6.20. All the three samples show a similar weak field dependence of the transport J_c in the whole field region. It should be noted that the transport J_c decreases with the increse of filament numbers, i.e. the reduction of filament size, so detailed microstructural characterizations are needed to explain the J_c degradation of 19- and 114-filament samples.

The optical micrographs of the longitudinal cross sections of polished 7-, 19- and 114-filament Sr-122/Ag/Fe tape are presented in Fig. 6.21. In Fig. 6.21a, it can be seen that the filament size of 7-filament Sr-122/Ag/Fe tape is very uniform in the tape axis direction. In Fig. 6.21b, in contrast to the uniform filament B1, filament B2 shows a wavy interface between the filament and sheath for the 19-filament Sr-122/Ag/Fe tape. The narrowed Sr-122 region marked with dashed box on filament will restrict the pass of larger currents. In 114-filament tape shown in Fig. 6.21c, though some uniform filaments like filament C1 can be observed, some others like filament C2 suffer from serious sausage effect due to the different hardness between

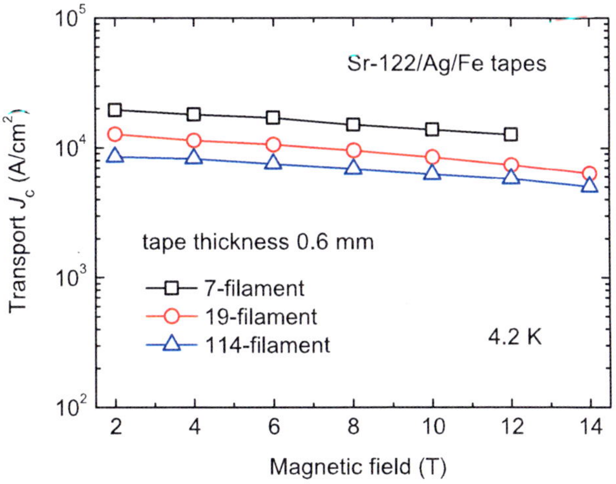

Fig. 6.20 Field dependence of the transport J_c for 7-, 19- and 114-filament Sr-122/Ag/Fe tapes of 0.6 mm in thickness. The applied field is parallel to the tape plane. Reprint from Yao et al. [16]

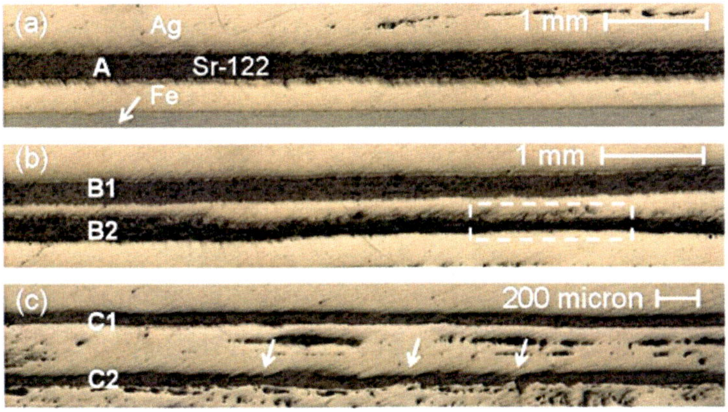

Fig. 6.21 Optical images of the polished longitudinal cross sections for **a** 7-, **b** 19- and **c** 114-filament Sr-122/Ag/Fe tape of 0.6 mm in thickness. The regions that restrict transport currents are marked with a dashed box and arrows. Reprint from Yao et al. [16]

the sheath and very fine filaments. As marked with arrows in filament C2, the necked regions will cause largely reduced or even discontinuous current path.

In order to characterize the micromorphology of the Sr-122 grains inside our multifilamentary tapes, 7-, 19- and 114-filament Sr-122/Ag/Fe tape of 0.6 mm in thickness were peeled off their iron sheath and part of silver matrix for SEM observation. In Fig. 6.22d, the inset shows a low-magnification SEM image of 114-filament tape after removing the sheath. It can be clearly seen that some residual Sr-122 filaments and silver matrix. The red rectangular box marks the area in which the high-magnification SEM image in the inset of Fig. 6.22c and the EDX spectrum in Fig. 6.22d were obtained. The EDX analysis comfirms that the area is consisted of Sr-122 phase. Similar to the inset of Fig. 6.22c, the insets of Fig. 6.22a, b show the micromorphology of Sr-122 phase for 7- and 19-filament tapes, respectively.

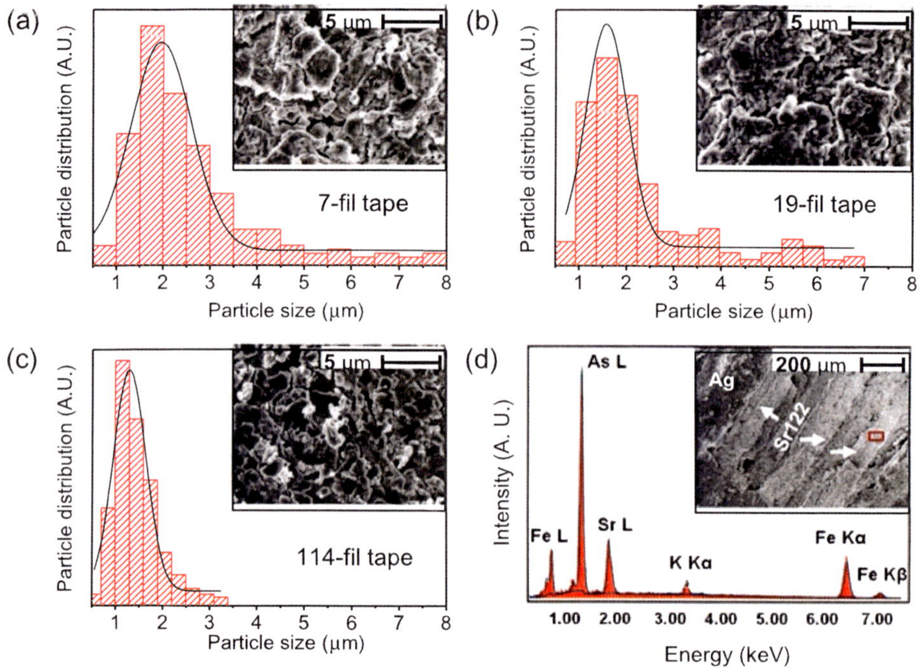

Fig. 6.22 Gauss fits of the particle size distribution (by measuring over 200 particles from SEM images) for the Sr-122 filaments in **a** 7-, **b** 19- and **c** 114-filament Sr-122/Ag/Fe tape of 0.6 mm in thickness. The insets show the typical SEM micromorphology of the Sr-122 filaments for the three tape samples. The inset of **d** is a SEM image showing the internal structure of a 114-filament tape after peeling off the iron sheath and part of silver matrix. The red rectangular box marks the area in which the magnified SEM observation for the inset of **c** and the EDX spectrum in **d** were performed. Reprint from Yao et al. [16]

Compared with the 7- and 19-filament tapes which contain inhomogeneous Sr-122 grains and their agglomerates, the 114-filament tape has many smaller Sr-122 grains. The distributions of their size and the corresponding Gauss fit curves for 7-, 19- and 114- filament tapes are plotted in Fig. 6.22a–c, and the peaks of the fitting curves are at 2.0, 1.6 and 1.3 μm, respectively. This result indicates that grain refinement occurs due to the crushing effect in deformation process with the reduction of filament size. For 114-filament tapes, Sr-122 particles larger than 4 μm were significantly eliminated. The refined grains can increase the density of grain boundaries and reduce the degree of grain texture, which are not beneficial to the J_c improvement, as for high-temperature cuprate superconductors [1]. Therefore, the J_c degradation in 19- and 114-filament Sr-122/Ag/Fe conductors can be ascirbed to the nonuniform filaments in longitudinal direction and the grain refinement effect.

According to the results presented above, there are still lagre room for the J_c improvement of multifilamentary 122 type iron pnictide conductors through three factors. Fisrt, in monofilamentary Sr(Ba)-122/Ag tapes, it has been proved that the transport J_c can be enhanced with the reducing of rolling thickness down to

0.30 ~ 0.26 mm [15]. A large transport J_c of 5.4 × 10^4 A cm^{-2} (4.2 K, 10 T) with a high average Vickers hardness of 136 in superconducting fialments has been reported in flat rolled thin Ba-122/Ag tapes of 0.3 mm in thickness [21], so we can expect thin multifilamentary tapes with higher transport J_c. However, due to the different hardness and workability of sheath materials, the deformation process suitable for such mechanically reinforced composites conductors is still needed to develop. Second, the sausage effect, which is a commonly observed phenomenon in BiSrCaCuO multifilamentary thin wires and tapes, can be alleviated by optimizing the reduction rate of deformation, the diameter of rollers, particle size of precursor, hardness of sheath materials and so on [22–25]. Third, the grain refinement in thin filaments can enhance grain boundary pinning but at the same time suppress intergrain currents. For MgB_2 superconductors, in which grain boundaries are transparent for current flow, the grain refinement has an overall positive effect to J_c performance [26, 27]. However, for iron pnictide superconductors, whose transport J_c can be suppressed by high-angle grain boundaries, we may try to further improve the grain connectivity by optimizing heat treatment. Therefore, the configuration and fabricating process for multifilamentary Sr-122/Ag/Fe composite conductors is still needed to improve. If high hardness, uniform interface and well-connected grains can be achieved for filaments, high transport J_c in multifilamentary iron-based superconductors can be achieved in the future.

6.5 Conclusion

Based on PIT process, Fe/Ag and Ag sheathed 122-IBS multifilamentary wires and tapes were fabricated by drawing and flat rolling process for the first time. The transport J_c of the Ag sheathed Sr-122 7- and 19-filament tapes can be enhanced to 6.1 × 10^4 and 3.5 × 10^4 A/cm^2 at 4.2 K and 10 T by hot press process, respectively. Such J_c values are the highest ever reported for IBS multifilamentary wires and tapes. However, considering the requirement of high mechanical strength and low cost for practical applications, the fabrication process for Fe/Ag sheathed 122-IBS multifilamentary wires and tapes were further investigated. 7- and 19-filament Sr-122 wires and tapes with silver as matrix and iron as outer reinforcing sheath were produced by the ex-situ PIT method. By hardness characterization, a positive correlation between hardness and transport J_c was revealed for 7- and 19-filament tapes. Due to the deformation induced densification, both 7- and 19-filament tapes show much higher hardness for Sr-122 filaments compared with round wires, but the mass redistributions of tapes exhibit inhomogeneity between different superconducting filaments. Based on six 19-filament Sr-122/Ag/Fe round wires, 114-filament Sr-122/Ag/Fe wires and tapes were also made by drawing and flat rolling process. Sr-122 filaments with size smaller than 50 μm were achieved in 114-filament round wires with 2.0 mm in diameter. The transport J_c for the 7-, 19- and 114-filament Sr-122 tapes were found to decrease with the increasing filament number, from 1.4 × 10^4 A/cm^2 for 7-filament tapes down to 6.3 × 10^3 A/cm^2 for 114-filament tapes at 4.2 K and 10 T. Nevertheless, all the transport J_c shows very weak magnetic field dependence at high fields, indicating a promising future of iron-based superconductors in practical applications. The degraded uniformity of interface between Sr-122 filaments and silver matrix, in addition to the grain refinement in very fine filaments were found to be the reason for the degradation of transport J_c in 19- and 114-filament Sr-122/Ag/Fe tapes. Further optimization for the process of multifilamentary Sr-122/Ag/Fe conductors is still needed to achieve filaments with high mass density and advantageous microstructure for improving the J_c-performance.

References

1. Larbalestier DC, Gurevich A, Feldmann DM, Polyanskii A (2001) High-T_c superconducting materials for electric power applications. Nature 414:368
2. Uglietti D (2019) A review of commercial high temperature superconducting materials for large magnets: from wires and tapes to cables and conductors. Supercond Sci Technol 32:053001
3. Sato K, Kobayashi S, Nakashima T (2012) Present status and future perspective of bismuth-based high-temperature superconducting wires realizing application systems. Jpn J Appl Phys 51:010006
4. Yao C, Ma Y, Zhang X, Wang D, Wang C, Lin H, Zhang Q (2013) Fabrication and transport properties of $Sr_{0.6}K_{0.4}Fe_2As_2$ multifilamentary superconducting wires. Appl Phys Lett 102:082602
5. Sasmal K, Lv B, Lorenz B, Guloy A, Chen F, Xue Y, Chu CW (2008) Superconducting Fe-based compounds $(A_{1-x}Sr_x)Fe_2As_2$ with A = K and Cs with transition temperatures up to 37 K. Phys Rev Lett 101:107007
6. Kovac P, Husek I, Pachla W (1997) Ceramic core density and homogeneity in BSCCO/Ag tapes. IEEE Trans Appl Supercond 7:2098
7. Han Z, Skov-Hansen P, Freltoft T (1997) The mechanical deformation of superconducting BiSrCaCuO/Ag composites. Supercond Sci Technol 10:371
8. Lee DF, Martin PM, Kroeger DM, Rupich MW, Li Q, Riley GN Jr (1997) Effects of initial cold work conditions on the deformation and current capacity of monofilamentary and multifilamentary Bi-2223 conductors. Supercond Sci Technol 10:702
9. Maeda H, Sastry PVPSS, Trociewitz UP, Schwartz J (2003) Critical current density and microstructures of thick monocore Bi2212 tapes grown in high magnetic fields. IEEE Trans Appl Supercond 13:3339
10. Beilin V, Dul'kin E, Yashchin E, Galstyan E, Lapides Y, Tsindlekht M, Felner I, Roth M (2004) Evolution of core connectivity in MgB_2 wires and tapes during PIT processing. Physica C 405:70
11. Wang C, Wang D, Zhang X, Yao C, Wang C, Ma Y, Oguro H, Awaji S, Watanabe K (2012) Improved J_c—B properties of MgB_2 multifilamentary wires and tapes. Supercond Sci Technol 25:125001
12. Husek I, Kovac P, Kopera L (1996) Study of BSCCO core density in multicore Ag sheathed tapes by microhardness profiles. Supercond Sci Technol 9:1066
13. Zeng R, Guo YC, Tanaka Y, Horvat J, Ionescu M, Beales TP, Apperley M, Liu HK, Dou SX (1999) Fabrication and properties of some Ag-alloy sheathed Bi-2223 tapes. IEEE Trans Appl Supercond 9:2710
14. Gao Z, Togano K, Matsumoto A, Kumakura H (2015) High transport J_c in magnetic fields up to 28 T of stainless steel/Ag double sheathed Ba122 tapes fabricated by scalable rolling process. Supercond Sci Technol 28:012001

15. Gao Z, Togano K, Matsumoto A, Kumakura H (2014) Achievement of practical level critical current densities in $Ba_{1-x}K_xFe_2As_2$/Ag tapes by conventional cold mechanical deformation. Sci Rep 4:04065
16. Yao C, Lin H, Zhang Q, Zhang X, Wang D, Dong C, Ma Y, Awaji S, Watanabe K (2015) Critical current density and microstructural homogeneity in multifilamentary $Sr_{1-x}K_xFe_2As_2$ composite conductors. J Appl Phys 118:203909
17. Weiss JD, Tarantini C, Jiang J, Kametani F, Polyanskii AA, Larbalestier DC, Hellstrom EE (2012) High intergrain critical current density in fine-grain $(Ba_{0.6}K_{0.4})Fe_2As_2$ wires and bulks. Nat Mater 11:682
18. Maiorov B, Katase T, Usov IO, Weigand M, Civale L, Hiramatsu H, Hosono H (2012) Competition and cooperation of pinning by extrinsic point-like defects and intrinsic strong columnar defects in $BaFe_2As_2$ thin films. Phys Rev B 86:094513
19. Fang L, Jia Y, Mishra V et al (2013) Huge critical current density and tailored superconducting anisotropy in $SmFeAsO_{0.8}F_{0.15}$ by low-density columnar-defect incorporation. Nat Commun 4:2655
20. Matsumoto A, Gao Z, Togano K, Kumakura H (2014) Critical current density enhancement of uniaxially pressed Ag sheathed (Sr, K)/Fe_2As_2 superconducting wires. Supercond Sci Technol 27:025011
21. Dong C, Yao C, Lin H, Zhang X, Zhang Q, Wang D, Ma Y, Oguro H, Awaji S, Watanabe K (2015) High critical current density in textured Ba-122/Ag tapes fabricated by a scalable rolling process. Scripta Mater 99:33
22. Riley GN, Gannon JJ, Miles PK, Parker DR (1994) Advances in the design and fabrication of high-T_c composite conductors. Appl Supercond 2:155
23. Wang WG, Horvat J, Liu HK, Dou SX (1996) The effect of mechanical deformation on silver-core interface and critical current density in Ag–Bi-2223 single- and multifilament tapes. Supercond Sci Technol 9:875
24. Zeng R, Beales TP, Liu HK, Dou SX (1998) Optimal reduction in rolling Ag-sheathed Bi-2223 multifilamentary tapes. Supercond Sci Technol 11:299
25. Sengupta S, Caprino E, Card K, Gaines JR, Motowidlo LR, Sokolowski RS, Garcia RR, Mukhopadyay SM (1999) Synthesis of bismuth-strontium-calcium-copper oxide powders for silver composite wires with uniform micron sized filaments. IEEE Trans Appl Supercond 9:2601
26. Malagoli A, Braccini V, Tropeano M et al (2008) Effect of grain refinement on enhancing critical current density and upper critical field in undoped MgB_2 ex situ tapes. J Appl Phys 104:103908
27. Malagoli A, Braccini V, Bernini C, Romano G, Vignolo M, Putti M, Ferdeghini C (2010) Study of the MgB_2 grain size role in ex situ multifilamentary wires with thin filaments Supercond. Sci Technol 23:025032

Chapter 7
Summary and Perspective

In summary, the very high upper critical field (H_{c2}) in addition to its low anisotropy makes IBS quite attractive for the construction of high-field magnets, which are desired for high-field applications such as next generation of MRI, NMR, accelerator, and fusion magnets. In this work, iron-based superconducting wires and tapes were fabricated with Ba-122 and Sr-122 precursors, various sheath configurations such as Ag, Fe and Fe/Ag composite, and different deformation processes based on ex-situ powder-in-tube (PIT) method. The superconducting properties and microstructure were systematically studied.

Novel methods including the co-addition of Ag and Pb metallic additives, the combined deformation process with rolling, intermediate annealing and cold pressing, and on-step hot pressing were proposed to improve the mass density for better grain connectivity and alleviate the weak-link effect at high-angle grain boundaries in the superconducting cores. A synergetic microstructural tailoring for mass density, grain alignment and micro-cracks is the key to realize high J_c performance for iron-based superconducting wires and tapes.

With MOI technique, magnetic flux distribution of the superconducting cores in Meissner state and remnant state was characterized, showing the remarkably improved grain connectivity in flat rolled Sr-122 tapes. With the HRTEM, lots of textured grains with clean and low-angle grain boundaries were observed in hot pressed Sr-122 tapes. It is found that the flux motion in Sr-122 tapes can be well explained by the thermal activated flux flow model, with which high pinning energy with weak field dependence is derived. The investigate for the mechanism of J_c enhancement for iron-based superconducting wires and tapes reveals that the high transport J_c achieved can be ascribed to the combined effect of good grain connectivity, high degree of texture, and strong pinning strength.

Based on the PIT method, Fe/Ag and Ag/Ag sheathed Sr-122 multifilamentary wires and tapes with filament numbers up 114 were fabricated for the first time. Due to the deformation induced densification, both 7- and 19-filament tapes show much higher hardness for Sr-122 filaments compared with round wires, but the mass

C. Yao, *Fabrication and Properties of High-Performance 122-Type Iron-Based Superconducting Wires and Tapes*, Springer Theses,
https://doi.org/10.1007/978-981-10-5184-5_7

redistributions of tapes exhibit inhomogeneity between different superconducting filaments. The transport J_c for the 7-, 19- and 114-filament Sr 122 tapes were found to decrease with the increasing filament number due to the degraded uniformity of Sr-122 filaments in the longitudinal direction, but all the transport J_c shows very weak magnetic field dependence at high fields, indicating a promising future of iron-based superconductors in practical applications.

For the future research and development of iron-based superconducting wires for practical applications, several issues are worthy of attention: (1) Further improvement on transport J_c: the high-J_c tapes processed with cold or hot press are still not in a complete optimization for mass density, grain texture and crack control. Therefore, it can be implied that there is still rooms for J_c improvement. (2) Engineering critical current density (J_e): for practical application J_e is usually concerned more by engineers. At present, the filling factor for Ag sheathed wires is about 20–25%, and for composite sheathed wires or multifilamentary wires, the filling factor is even lower, so improving J_e by reducing the use of metal sheath is also important for practical applications. (3) Optimization for the fabrication of composite sheathed wires: for wires with stiff metal as outer reinforcing sheath and Ag as inner barrier sheath, uniform deformation is still challenging in cold-work process. Systematical optimizations for the cold work parameters for such composite sheathed wires combined with appropriate intermediate annealing are needed. (4) Detailed investigation for mechanical and electromagnetic properties: at present the mechanical properties are not fully understood for iron-based superconducting wires and tapes, and investigations for AC loss and thermal stability are also needs to be conducted for wires and tapes with different configurations and architectures.